Interfacing PIC Microcontrollers to Peripherial Devices

International Series on

INTELLIGENT SYSTEMS, CONTROL, AND AUTOMATION: SCIENCE AND ENGINEERING

VOLUME 49

For other titles published in this series, go to
www.springer.com/series/6259

Bohdan Borowik

Interfacing PIC Microcontrollers to Peripherial Devices

Springer

Bohdan Borowik
Akademia Techniczno-Humanistyczna
Bielsko-Biala
Katedra Elektrotechniki i
Automatyki
Willowa 2
43-309 Bielsko-Biala
Poland
bo@borowik.info

ISBN 978-94-007-1118-1 e-ISBN 978-94-007-1119-8
DOI 10.1007/978-94-007-1119-8
Springer Dordrecht Heidelberg London New York

Library of Congress Control Number: 2011921707

Cover design: SPI Publisher Services

Printed on acid-free paper

Springer is part of Springer Science+Business Media (www.springer.com)

Table of contents

Preface

Our book is targeted for students of electronics and computer sciences. First part of the book contains 15 original applications working on the PIC microcontroller. They are: lighting diodes, communication with RS232 (bit-banging), interfacing to 7-segment and LCD displays, interfacing to matrix keypad 3 x 4, working with PWM module and other. They cover 1 semester teaching of microcontroller programming or similar clases. The book has schematics diagrams and source codes in assembly with their detailed description.
All tests were prepared on the basis of the original documentation (data sheets, application notes). Sometimes, encountering problems we looked for help on various foums in the world with people involved in the hi tech challenges.

Next three chapters: The Stack, Tables and Table instruction and Data memory pertains to PIC18F1320. Software reffered to is also in assembly laguage.

Finally we describe the application of the PIC24FJ microcontroller with the 240x128 LCD display and the analog accelerometer sensor.

Testing board description

Presented in the book applications were implemented on the original testing board called Microcon4. The hardware is uncomplicated and showing parts of entire schematics is intended to illustrate the easy of use of various peripherial devices. We use following peripherial devices:

- ICSP In-Circuit-Serial Programming device
- 7-segment display
- TTL/CMOS driver ULN 2803 for Port A and Port B
- LCD display
- Matrix Keypad
- I2C expander PCF 8574
- EEPROM 24C02 and RTC PCF8583
- UART communication bus with MAX 232 IC

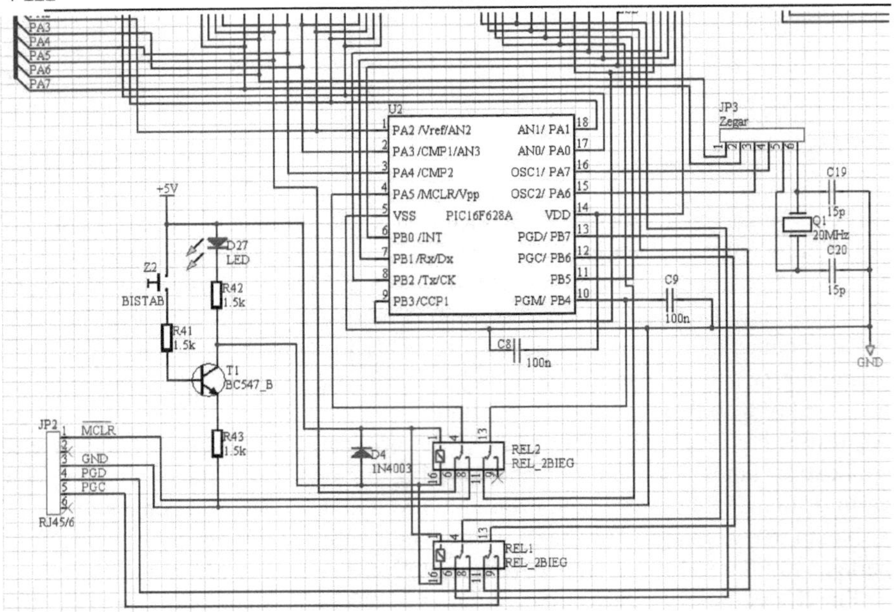

Fig. 1. In Circuit Serial Programing ICSP device connected to JP2

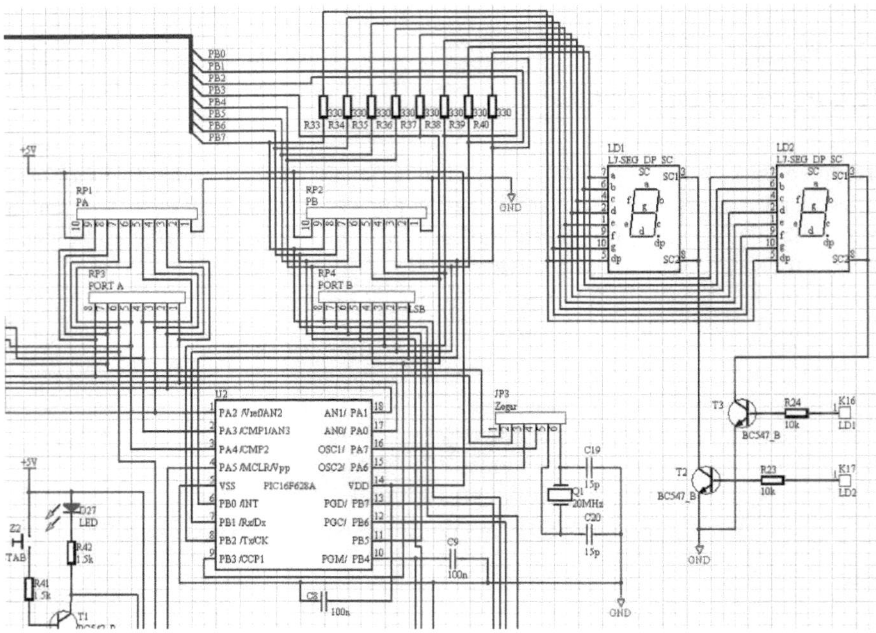

Fig. 2. 7-Segment Display connected to port B

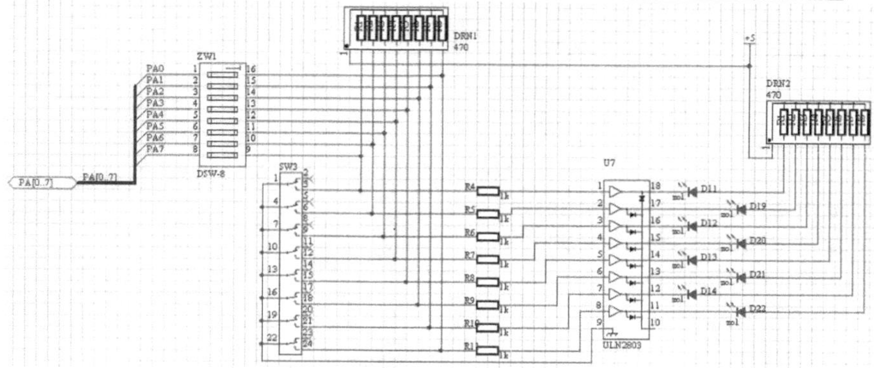

Fig. 3. Using TTL/CMOS driver ULN 2803 for port A

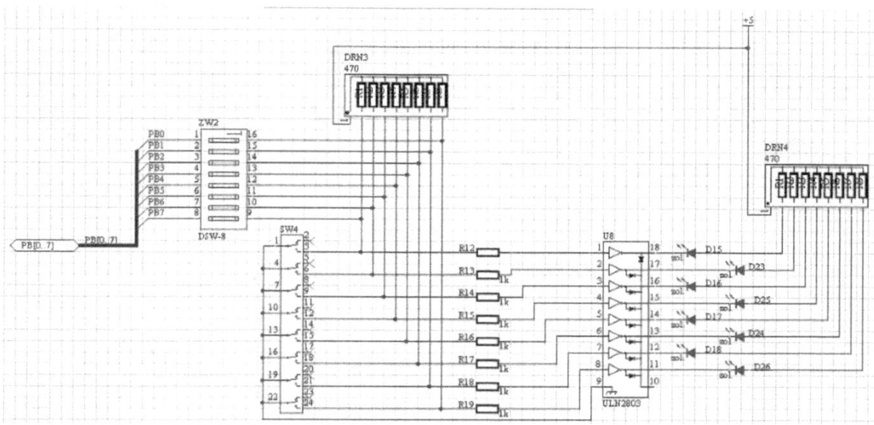

Fig. 4. Using TTL/CMOS driver ULN 2803 for port B

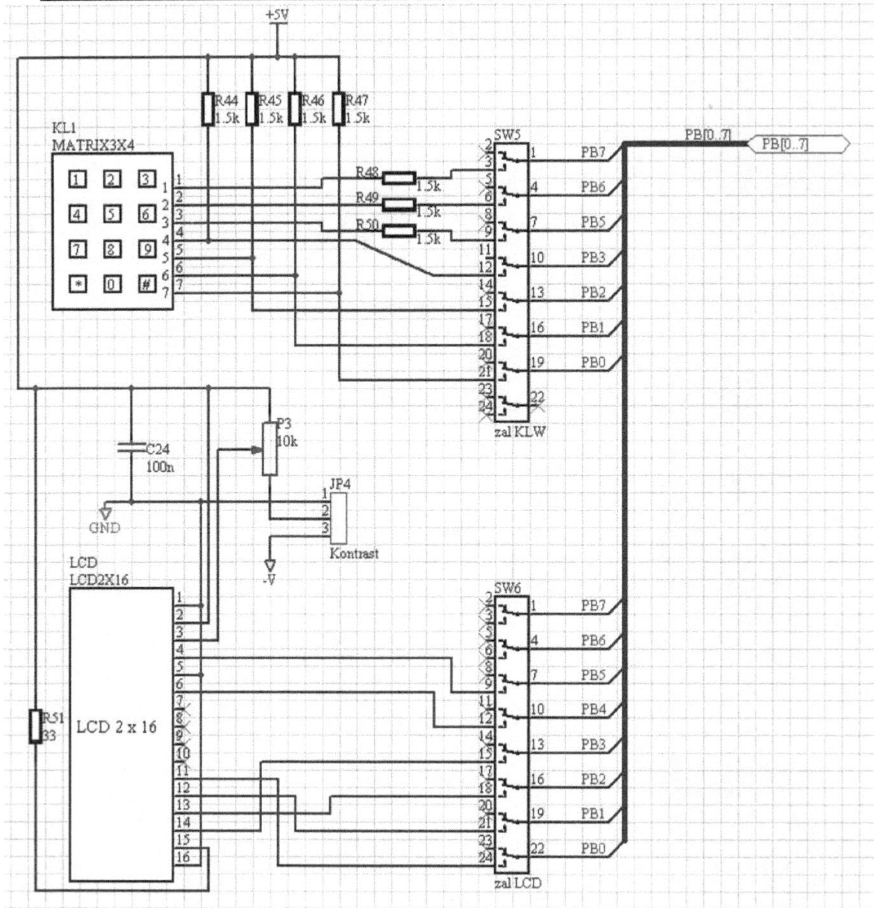

Fig. 5. Connection LCD Display and Matrix keyboard to port B

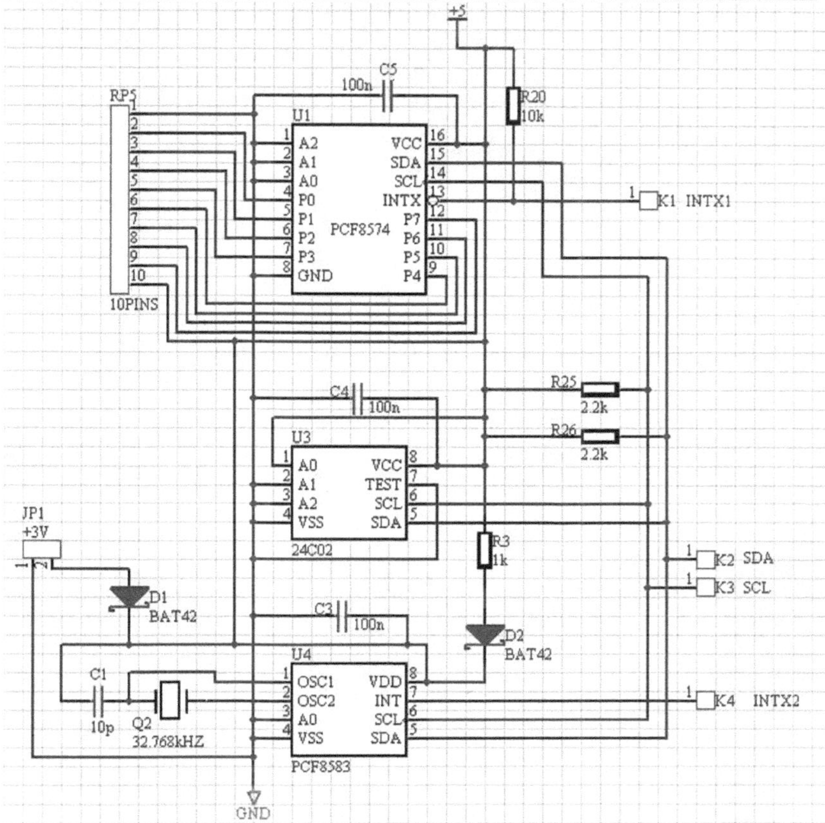

Fig. 6. Connection Expander I2C, PCF8574 EEPROM 24C02 and RTC PCF8583

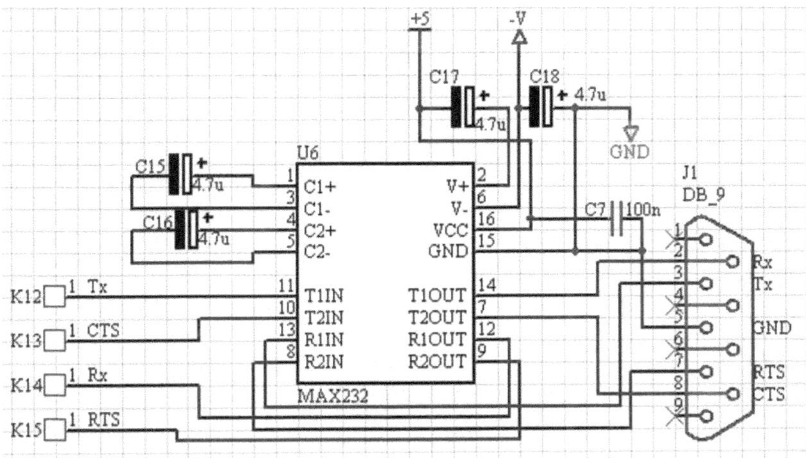

Fig. 7 UART communication circuit apply MAX232

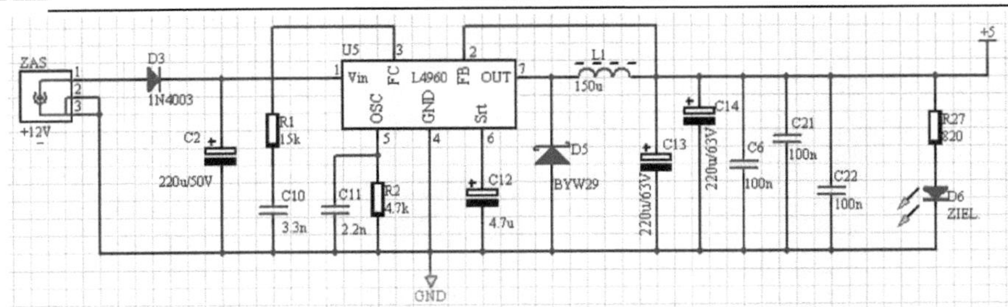

Fig. 8. Switching power supply using the U5 L4960 device.

Additionaly we present the schematic for typical programmer:

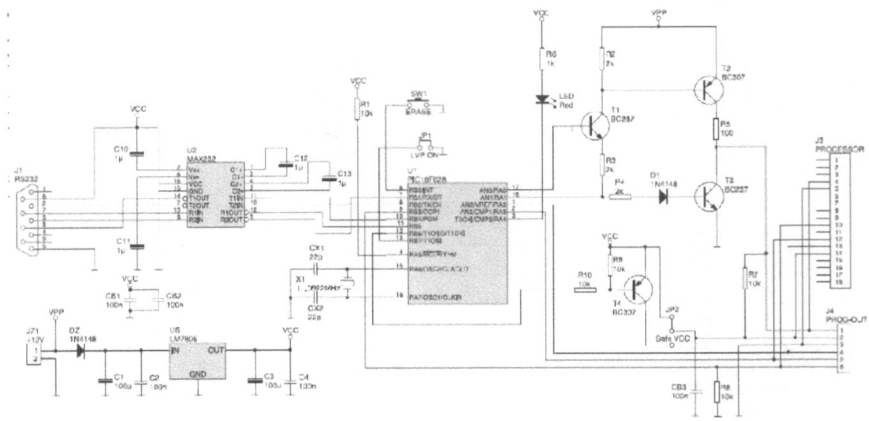

Fig. 9. Schematic diagram of ICSP programmer

Test examples

Presented below tests were implemented on the testing board Microcon4.

Test 1. Turn the LED on for the calculated period of time.

We define two registers *count1* and *count2* in the data memory GPR (General Purpose Registers), that starts at the address 0x20.

```
cblock    h'20'    ; constant block
    count1, count2
endc
```

In the nested loop the inner loop runs 256 times, therefore count1 is first cleared and then is decremented 256 times, until will have again value of 0.

```
        decfsz      count1, f
```

The outer loop will run 9 times, therefore we initiate register count2 with the value of 9.

The loop requires 3 machine cycles

```
del decfsz      count1, f           ; inner loop
    goto  del
```

Decrementation execute in a single cycle, while program branch requires 2 cycles.For the internal clock frequency of 32 kHz, 1 machine cycle lasts 108 μs. Calculated time:

9 x 256 x 3 x 108 μs = 0,746 s.

Program code:

```
; ******************************************
;T1 Blinking the LED, connected to bit1 of PORT B,
```

B. Borowik, *Interfacing PIC Microcontrollers to Peripherial Devices*, Intelligent Systems, Control and Automation: Science and Engineering 49, DOI 10.1007/978-94-007-1119-8_1,

```
;    with delay equal to 0,75 s.
;    Internal clock frequency 37 kHz, Tcm = 108 µs
; ******************************************

    list p=16f628, r=hex     ; declare processor,
                             ; specifying the radix
    #include p16f628.inc   ; include register label
                             ;  definitions
    __config h'3f10'         ; configuration
                             ; information
                             ; for selected processor
    errorlevel -302          ; turn off banking
                             ; message
#define out1    portb,1      ; the directive
                        ; substitutes a text  portb,1
                             ; with a string out1

#define count0  .9      ;defines value of 9 decimal
                        ; for register count2
cblock    h'20'    ; constant block
    count1, count2
endc
    movlw h'07'         ;07 -> w
    movwf cmcon         ; w->cmcon, comparators off
    clrf  porta         ; clear PORTA output latches
    clrf  portb         ;initializes PORTB
    bsf   status, rp0 ;bank 1
    bcf   pcon, oscf          ;internal gen.32 kHz,
                              ; cm=108µs
    clrf  trisa         ;PORTA for output
    clrf  trisb         ;PORTB for output
```

```
    bcf   status, rp0                ;bank 0
loop                           ; main loop
    bsf   out1          ; turn the LED on
    call  delay         ; call delay routine
    bcf   out1          ; turn off LED on RB1
    call  delay         ; call delay routine
    goto loop           ; repeat main loop

delay            ;delay 0,75s (9*256*3*108µs=0.746 s)
    movlw count0               ; count0 ->w
    movwf count2               ; initializes count2
    clrf  count1               ; initializes count1

del decfsz      count1, f           ; decrement count1
    goto  del                  ; if not 0, keep
                               ; decrementing count1
    decfsz      count2, f      ; decrement count2
    goto        del            ; if not 0, decrement
                               ; count1 again
    return                     ; return to main routine
end
; ***************************************************
```

Test 2. Turn on the LEDs connected to various lines of port B

Program code:

```
; ***********************************************
; * T2 *  Turning on the LEDs, connected to
;    bit 1 and bit 7 of PORT B by seting  RA1
;    and RA7 to high.
;    Internal clock frequency 37 kHz, Tcm = 108 µs
; ***********************************************

    list p=16f628, r=hex      ; declare processor,
                              ; specyfying the radix
    #include p16f628.inc  ; include register label
                              ; definitions
    __config h'3f10'          ; configuration
                              ; information
                              ; for selected processor
    errorlevel -302           ; turn off banking
                              ; message

    movlw h'07'       ;07 -> w
    movwf cmcon       ; w->cmcon, comparators off
    clrf  porta       ; clear PORTA output latches
    clrf  portb       ;initializes PORTB
    bsf   status, rp0 ;bank 1
    bcf   pcon, oscf          ;internal gen.32 kHz,
                              ; Tcm=108µs
    clrf  trisa       ;PORTA for output
    clrf  trisb       ;PORTB for output

    bcf   status, rp0                ;bank 0
```

B. Borowik, *Interfacing PIC Microcontrollers to Peripherial Devices*, Intelligent Systems, Control and Automation: Science and Engineering 49, DOI 10.1007/978-94-007-1119-8_2,

```
      bsf portb, 0              ; LED 0 on
      bsf portb, 7              ; LED 7 on
      goto $                    ; go to self
                                ; loop here forever
  end
  ; ************************************************
```

Note:
The LED on RA5 is turned on despite of initializing port A and port B with 0x00:

```
      clrf  porta        ; clear PORTA output latches
      clrf  portb        ;initializes PORTB
```

because it is ~MCLR line.
RB0 and EB7 lines are set to high with the instructions:

```
      bsf portb, 0              ; LED 0 on
      bsf portb, 7              ; LED 7 on
```

Another way for turning selected LEDs on will be copying the bitmap mask to portB. It will be subject of test 3.

Test 3. Turn on the LEDs connected to various lines of port B

(continued)

Like Test 2, but the mask for the whole PORTB is used instead of setting up particular lines. The mask for seting lines 0 and 7 would be:

```
movlw b'10000001'
movwf portb
```

```
; ************************************************
; * T3 *     Turning on the LEDs, connected to
;    bit 1 and bit 7 of PORT B by seting  RA1
;    and RA7 to high.
*******************************************
    list p=16f628, r=hex     ; declare processor,
                             ; specyfying the radix
    #include p16f628.inc   ; include register label
                             ; definitions
    __config h'3f10'         ; configuration
                             ; information
                             ; for selected processor
    errorlevel -302          ; turn off banking
                             ; message

    movlw h'07'       ;07 -> w
    movwf cmcon       ; w->cmcon, comparators off
    bsf   status, rp0 ;bank 1
    bcf   pcon, oscf         ;internal gen.32 kHz,
                             ; Tcm=108µs
    clrf  trisa       ;PORTA for output
    clrf  trisb       ;PORTB for output

    bcf status, rp0          ;bank 0
```

B. Borowik, *Interfacing PIC Microcontrollers to Peripherial Devices*, Intelligent Systems, Control and Automation: Science and Engineering 49, DOI 10.1007/978-94-007-1119-8_3,

```
    clrf  porta         ; clear PORTA output latches
    clrf  portb         ;initializes PORTB
    movlw b'10000001'         ; w <- b'10000001'
    movwf portb               ; portb <- w
    goto $                    ; go to self
    end
```

Test 4. All LEDs connected to portb blink with different frequencies

For the internal generator frequency of 37 kHz the machine cycle equals 108 us. In order to achieve LEDs blinking with various frequency we use TMR0 prescaler. Frequencies are deployed within the range: 55 ms to 7.079 s according to the table:

Table . Table 1. Various times of blinking LED for particular lines of portb

Bit	Time calculation for TMR0 register	Blinking time
0	256 * 2 * T_{CM}	55 ms
1	256 * 4 * T_{CM}	110 ms
2	256 * 8 * T_{CM}	221 ms
3	256 * 16 * T_{CM}	442 ms
4	256 * 32 * T_{CM}	885 ms
5	256 * 64 * T_{CM}	1.769 s
6	256 * 128 * T_{CM}	3.538 s
7	256 * 256 * T_{CM}	7.079 s

Prescaler value of 256 is obtained by setting up control bits PS2, PS1 and PS0 in the OPTION_REG register as follows:

```
      movlw  b'10000111'        ;  preskaler 256
      movwf option_reg  ;PSA=0 t0cs=Tcm presk. dla TMR0
                        ; prsk=256
```

During each iteration the contents of TMR0 counter is copied to PORTB>

```
   loop                         ; main loop
            movf tmr0, w        ;
            movwf portb
            goto loop
```

All portb lines are tested, It gives the effect of lighting cascade.

```
  ; ************************************************
  ; * T4 *  Blinking LEDs on portb with different fre-
quencies
  ;     TMR0 determines the time for particular lines.
  ;     internal clock frequency 37 kHz, Tcm = 108 us
  ; ************************************************
     list p=16f628, r=hex     ; declare processor,
```

B. Borowik, *Interfacing PIC Microcontrollers to Peripherial Devices*, Intelligent Systems, Control and Automation: Science and Engineering 49, DOI 10.1007/978-94-007-1119-8_4,

```
                               ; specyfying the radix
    #include p16f628.inc        ; include register label
                               ; definitions
    __config h'3f10'           ; configuration
                               ; information
                               ; for selected processor
    errorlevel -302            ; turn off banking
                               ; message

    movlw h'07'          ;07 -> w
    movwf cmcon          ; w->cmcon, comparators off
    clrf  porta          ; clear PORTA output latches
    clrf  portb          ; initializes PORTB

    bsf   status, rp0 ;bank 1
    bcf   pcon, oscf           ;internal gen.32 kHz,
                               ; Tcm=108µs
    clrf  trisa          ;PORTA for output
    clrf  trisb          ;PORTB for output

    ; contents option_reg
    movlw  b'10000111'         ;  prescaler 256
    movwf option_reg   ;PSA=0, t0cs=Tcm, presc. For
                       ; TMR0 =256
    bcf status, rp0                   ;bank 0

loop                           ; main loop
     movf tmr0, w       ; check value of the subsequent
                        ; bit
     movwf portb
     goto loop
    end
```

Test 5. Acoustic signal of 1 kHz frequency generated with PWM module

```
; ************************************************
; * T5 * Acoustic signal received from PWM module.
;        Time measured by TMR1 counter
;        PWM generated signal frequency: 1 kHz
;        outputed on RB3
;     internal clock frequency 37 kHz, Tcm = 108 us
; ************************************************
    list p=16f628, r=hex     ; declare processor,
                             ; specifying the radix
    #include p16f628.inc    ; include register label
                             ; definitions
    __config h'3f10'         ; configuration
                             ; information
                             ; for selected processor
    errorlevel -302          ; turn off banking
                             ; message
#define lpr2 .249            ; final value of the tmr2
                             ; counter
     movlw h'07'             ;07 -> w
     movwf cmcon       ;w->cmcon, comparators off
     movlw b'00000010'
     movwf porta             ; initialize PORTA
     clrf portb              ; initialize PORTB
     clrf tmr1l              ; clear TMR1L
     clrf tmr1h              ; clear TMR1H
     clrf pir1         ; clear pir1 (tmrlif flag)
     bsf status, rp0         ;bank 1
     clrf trisa              ;port a output
     clrf trisb              ;port b output
```

B. Borowik, *Interfacing PIC Microcontrollers to Peripherial Devices*, Intelligent Systems, Control and Automation: Science and Engineering 49, DOI 10.1007/978-94-007-1119-8_5,

```
        ; initialise pr2 register
    movlw lpr2              ;lpr2->w
    movwf pr2               ;pr2 <- w
    bcf status, rp0         ;bank 0
    movlw .125              ; PWM Duty Cycle 50%
    movwf ccpr1l
   movlw b'00000101'        ; tmr2 enabling
                            ; and configuring
   movwf t2con              ; prescaler=4,
                            ; postscaler=1
   movlw b'00001100' ;CCP configuring
   movwf ccp1con        ; PWM mode
    movlw b'00110001'       ;configuration, tmr1 on
     movwf t1con            ;tmr1cs=Tcm, prescaler=8
          ;overflow after  256*256 *8 *Tcm =.52 s

loop                    ; main loop
    bsf status, rp0               ; bank 1
      movlw b'00001000'           ; w <- b'00001000'
      xorwf trisb, f              ; toggling the
                                  ; buzzer on RB3
    movlw b'00000010'             ; w <- b'00000010'
    xorrwf trisa, f               ; toggling LED RA1
    bcf status, rp0         ; bank 0
spr      btfss pir1, tmr1if       ; test tmr1if flag
    goto spr                ; wait for setting
                            ; the flag up
    bcf  pir1, tmr1if       ; clear the flag
    goto loop
    end
   ;
************************************************
```

Program description

Configuring the CCP module for PWM operation requires the following steps:

1. Setting the input signal frequency.

$$T_{PWM} = (PR2 + 1) . Tcm . T2_{prescaler}$$

$$1/1 \text{ kHz} = 0{,}001 \text{ s} = 1000 \text{ µs}$$

$$Tcm = 1 \text{ µs (for f osc.} = 4 \text{ MHz)}$$

T_{PWM} is specified by the value of the PR2 register, the clock oscillation period and the prescaler value.

T_{PWM} = 1000 µs matches two values: 250, 4. Let's assume prescaler value = 4 and PR2 = 256.

```
...
#define lpr2 .249      ;   final   value   of   the   tmr2
counter
...
      movlw lpr2  ; w <- lpr2
      movwf pr2         ; pr2 <- w
...
      movlw b'00000101' ; tmr2 enabling and configuring
      movwf t2con       ; prescaler=4, postscaler=1
```

T2CON register enables or disables the timer and configures the prescaler and postscaler.

Control bits 1-0 equal 01	prescaler = 4,
Bit 2 equal 1	Timer2 is on,
Bits 3-6 equal 0000	postscaler = 1.

2. Assume PWM Duty Cycle equal to 50%, or 0,5.

We set this 10 bit value by writing to the CCPR1L register and DC1M1 and DC1M0 bits of CCP1CON register using the following formula:

PWM Duty Cycle = (4 x CCPR1L + 2 x DC1M1 + DC1M0) / 4 x (PR2 + 1)

therefore

0,5 x 4 x (PR2 + 1) = (4 x CCPR1L + 2 x DC1M1 + DC1M0)

For PR2 = 249 on the left side we obtain 500. In 10 bit binary it is:

500 = b'0111 1101 00'

Bits are grouped because upper 8 bits represent CCPR1L register and are equal: 125 (b'0111 1101').

Lower 2 bits are equal to 0. Those values we have to write to DC1M1 and DC1M0 bits of the CCP1CON register.

```
...
        movlw .125          ;50% PWM Duty Cycle
        movwf ccpr1l
...
        movlw b'00001100' ;CCP configuring
        movwf ccp1con         ; PWM mode
```

DC1M1 and DC1M0 are bits CCP1CON<5:4>.
Bits 0-3 equal to 1100 set PWM mode.
Some instructions refer to Timer1 module operation.

```
        ...
        clrf    tmr1l
        clrf    tmr1h
        ...
        clrf    pir1
        ...
        movlw         b'0011 0001'
        movwf         t1con
```

Timer1 is used for toggling state on lines RB3 (buzzer) and RA1 (LED). Toggling time is calculated from the formula:
256 * 256 * 8 * Tcm = .52 s
because for Timer1 prescaler is set to 8, and TMR1IF flag sets its value after whole 16 bits register (TMR1L and TMT1H) overflows.

```
spr        btfss pir1, tmr1if        ; check tmr1if
      goto spr                       ; wait for setting
                                     ; the flag up
```

When TMR1 register overflows, the flag is set and TMR1 register is reset to 0. Then the TMR1IF flag need to be software cleared:

```
        bcf   pir1, tmr1if   ;clear the flag
```

As was shown, the value of 0011 0001 was written to T1CON register.
Bits <5:4> (11) set prescaler to 8.
At the beginning of the program the TMR1IF flag of the PIR1 register is also cleared by setting whole register to 0.

```
clrf    pir1
```

Schematic below shows connection of the piezoelectric buzzer to the module.

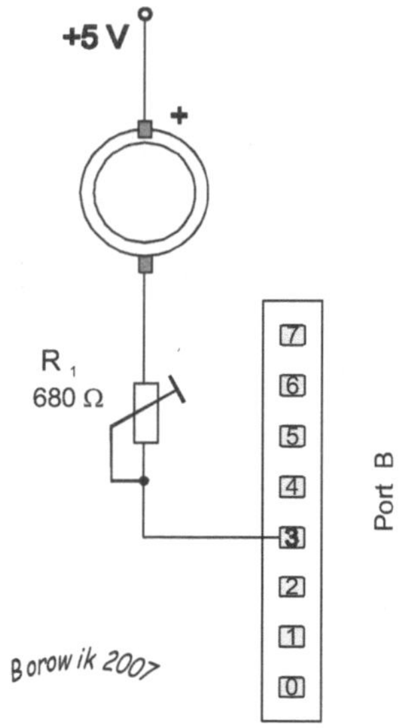

Fig. 10. Connection of the piezoelectric buzzer to the module.

Test 6. Morse code: PIC generated with PWM

Generating the word PIC in Morse code using PWM module.
Generating acoustic signal with PWM was described in previous test.

```
; ***********************************************
; T6 Morse code. Acoustic signal is received from
;    PWM module; Buzzer is connected to RB2
;    internal gen. 4 MHz; Tcm = 1 us
; ***********************************************
    list p=16f628, r=hex
    #include p16f628.inc       ; include register label
                               ; definitions
    __config h'3f10'           ; configuration
                               ; information
                               ; for selected processor
    errorlevel -302            ; turn off banking
                               ; message
  cblock  h'20'          ; constant block
      l2,l3
  endc
 #define lop .255
 #define lpr2 .243

            ; beginning of the program
    movlw h'07'                ; 07 -> w
    movwf cmcon          ; w->cmcon, comparators off
    clrf portb                 ; initialize PORTB
    bsf status, rp0            ; bank 1
    clrf trisb                 ; port b output
    bsf   pcon, oscf           ; internal gen. 4 MHz,
                               ; Tcm = 1 us
    movlw lpr2
    movwf pr2
```

B. Borowik, *Interfacing PIC Microcontrollers to Peripherial Devices*, Intelligent Systems, Control and Automation: Science and Engineering 49, DOI 10.1007/978-94-007-1119-8_6,

```
    bcf status, rp0          ; bank 0
    movlw .125       ;50% PWM Duty Cycle
    movwf ccpr1l
    movlw b'00000101'
    movwf t2con
    movlw b'00001100'
    movwf ccp1con
    bsf status, rp0          ;  bank 1
    bsf trisb,3              ; buzzer off
    bcf status, rp0

loop          ; main loop
        nop     ; wait, stabilize
        nop     ; wait, stabilize
    ;  P I C   . - - .      . .    - . -

    call przerwa
    call przerwa
    call przerwa
    call przerwa

    call dot
    call przerwa
    call dash
    call przerwa
    call dash
    call przerwa
    call dot

    call przerwa
    call przerwa
    call przerwa
```

```
    call dot
    call przerwa
    call dot

    call przerwa
    call przerwa
    call przerwa

    call dash
    call przerwa
    call dot
    call przerwa
    call dash

    goto loop

          ;  routines
dot
    bsf status, rp0
    bcf trisb,3        ; buzzer on
    bcf status, rp0
    call del           ; dot
    bsf status, rp0
    bsf trisb,3        ; buzzer off
    bcf status, rp0
    call del           ; wait
    return

dash                  ; dash = 3 *  dot
    bsf status, rp0
    bcf trisb,3        ; buzzer on
```

```
    bcf status, rp0
    call del            ; dash
    call del            ; dash
    call del            ; dash
    bsf status, rp0
    bsf trisb,3         ; buzzer off
    bcf status, rp0
    call del            ; wait
       return

przerwa                           ; delay
    call del
    call del
    return

del          ; how long last dot and space
             ; 255 x (256 x 4 cycles x 1us)=ca 0.261 s
    movlw    lop  ; w <- .255
    movwf    13    ; 13 <- w
    clrf     12    ; clear 12

spr
    nop          ; extending delay of the inner loop
    decfsz 12, f
    goto spr     ; inner loop
    decfsz 13, f
    goto spr     ; outer loop
    return
      end
```

Program description

Routine *del* contains nested loop. Two general purpose registers 12 and 13 are decremented. 12 is first cleared and then is decremented in the inner loop with instruction decfsz (decrement f, skip if zero). The loop is executed 256 times. It takes time of 1.024 ms (256 x 4 x 1us).
13 user register is decremented 255 times.
Loop requires about 0.261 s (255 x 1.024 ms) to execute.
Nop instruction adds up to the number of machine cycles in one iteration.

```
spr nop                  ; 1 cycle
    decfsz 12, f         ; 1 cycle
    goto spr             ; 2 cycles
```

Total 4 cycles in the inner loop.

Test 7. LED turn on after pressing switch on RB4

Pressing switch on RB4 toggles LED on RA1.

```
; ***********************************************
; T7 Pressing switch on RB4 causes testing the state
; of the line RA1 and changing its value: 0->1, 1->0
; internal gen.32 kHz, Tcm=108µs
; ***********************************************
    list p=16f628, r=hex       ; declare processor,
                               ; specifying the radix
    #include p16f628.inc       ; include register label
                               ; definitions
    __config h'3f10'           ; configuration
                               ; information
                               ; for selected processor
    errorlevel -302            ; turn off banking
                               ; message
licz equ h'21'
#define op .150                ; value for setting delay
#define ou      porta, 1        ; LED on RA1
#define in      portb, 4        ; switch on RB4

   movlw h'07'          ; 07 -> w
    movwf cmcon         ; w->cmcon, comparators off
    clrf porta                ; initialize PORTA
   bsf status, rp0      ; bank 1
   bcf    pcon, oscf          ;internal gen.32 kHz,
                              ; Tcm=108µs
   clrf trisa           ; porta output
   bcf option_reg, not_rbpu        ; PORTB pull-ups
                                   ; are enabled
   bcf status, rp0                 ;bank 0
```

B. Borowik, *Interfacing PIC Microcontrollers to Peripherial Devices*, Intelligent Systems, Control and Automation: Science and Engineering 49, DOI 10.1007/978-94-007-1119-8_7,

```
                        ; routines
lin4
   btfss  in  ; testing switch on RB4
   goto zapal          ; turn LED on
   goto zgas           ; turn LED off

zapal
   call opoz           ; delay
   bsf ou              ; RA1 (LED)high
   goto lin4

zgas
   call opoz
   bcf ou              ; RA1 (LED)low
   goto lin4

opoz                         ; delay (t = Tcm*3*op)
   movlw   op          ; w <- op
   movwf   licz        ; licz <- w
   decfsz licz,f       ; decrement counter licz
   goto $ - 1          ; decrement again
   return

   end
```

In the *lin4* loop the RB$ line status is constantly checked.

```
lin4
   btfss  in           ; Is switch pressed?
   goto zapal   ; LED on
   goto zgas           ; LED off
```

In the loop two routines are called: turning LED on and off. Both routines call also routine for delaying, generating delay of 48.6 ms:

$$Tcm \times 3 \text{ cycles} \times op = 108 \text{ us} \times 3 \times 150 = 48.6 \text{ ms}$$

On the picture below on the evaluation board, at the right down side there are two 8 bit ports: port A and above port B. The pencil points to port B. All port B lines are on HIGH, except of line RB4. Diodes attached to port B are lighting, but LED attached to RB$ is off, because the switch connected to RB4 is pressed down.
After pressing any of RB4 – RB7 switches the LED on RA1 get lighted.
If no switch is depressed (high on particular port B line) then, in the *lin4* loop, after the conditional instruction:

```
lin4
    btfss   in              ; testing switch on RB4
    goto zapal              ; turn LED on
    goto zgas               ; turn LED off
```

the next instruction `(goto zapal)` is omitted. Forcing low on RB4 causes executing next instruction and turning LED on RA1 on.
We see two LEDs lighting on PORT A: on RA5, which is input only port, always kept high. When this pin is configured as ~MCLR, is an active low Reset to the device. LED connected to RA1 lights when switch on RB4 is pressed.

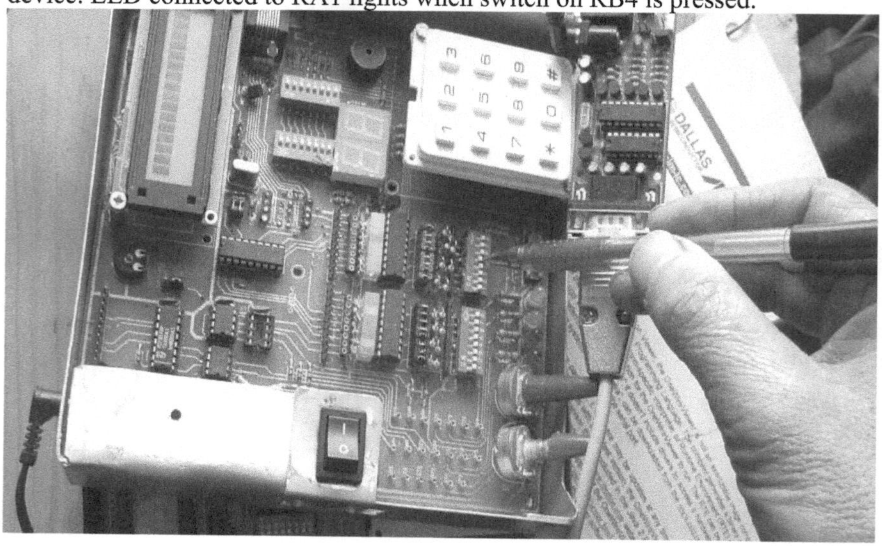

Fig. 11. Switching RB4 to low causes turning on LED on RA1

Test 8. Waking the device from SLEEP with RB4 interrupt-on-change

Test 8 ilustrates RB4 interrupt-on-change feature, that wakes up the controller from the sleep. Then LED is turned on for a 0,4 s.

```
; ************************************************
;T7 After pressing the switch on RB4 interrupt occurs
;and wake processor from sleep. Then the program
; continue execution: turn the LED on for a time of
; 0.1 s and again the processor is put into
; SLEEP mode.
; internal gen.32 kHz, Tcm=108µs
; ************************************************
    list p=16f628, r=hex      ; declare processor,
                              ; specifying the radix
    #include p16f628.inc      ; include register label
                              ; definitions
    __config h'3f10'          ; configuration
                              ; information
                              ; for selected processor
    errorlevel -302           ; turn off banking
                              ; message

#define wy  porta, 1  ; LED on RA1
#define we  portb, 4  ; switch on RB4
l1 equ h'20'                  ; assign Variable_Name
                              ; to Data Memory address

      ; begining of the program
   movlw h'07'                ;07 -> w
   movwf cmcon              ; w->cmcon, comparators off
   clrf  porta                ; initialize PORTA
   bsf status, rp0            ; bank 1
```

B. Borowik, *Interfacing PIC Microcontrollers to Peripherial Devices*, Intelligent Systems, Control and Automation: Science and Engineering 49, DOI 10.1007/978-94-007-1119-8_8,

```
    clrf trisa                  ; set port a as output
    clrf option_reg             ; clear option_reg,
                          ; PORTB pull-ups are enabled
    bcf pcon, oscf              ; internal gen.32 kHz,
                                ; Tcm=108μs
    bcf status, rp0             ; bank 0
    clrf l1                     ; clear l1
    movf portb, f               ; initialize port b
    clrf intcon     ; clear intcon, clear rbif flag
    bsf intcon, rbie            ; set rbie mask

  loop      ; main loop
     sleep                      ; sleep mode
     bsf wy                     ; turn the LED on
     call del                   ; delay .1 s
     bcf wy                     ; turn the LED off
     call del
     call del
     call del
     call del
     call del

    btfss  we             ; Is the switch released?
    goto $-1
    bcf intcon, rbif      : clear rbif flag
    goto loop   ;
del  nop
    nop
    nop
    nop
    decfsz l1, f  ; decrement l1
    goto del
```

```
    return


    end
;   *********************************************************
```

Program description

Four PORTB pins, RB<7:4>, if configured as inputs, have an interrupt-on-change feature. This interrupt sets flag bit RBIF and can wake the device from SLEEP. Interrupt is enabled after setting the RBIE mask:

```
        bsf intcon, rbie
```

Interrupt on mismatch feature together with software configurable pull-upps on these pins allow easy interface to a switch and make it possible for wake up on switch depression.

After clearing option_reg register, PORTB pull-ups are enabled.

```
    clrf option_reg    ; clear option_reg,

                       ; PORTB pull-ups are enabled
```

Instruction `movf portb, f` initializes PORTB, because any read or write of PORTB will end the mismatch condition and allow flag bit RBIF to be cleared.

After `sleep` instruction the device enters the sleep mode and waits for interrupt to be waked up.

If the interrupt is enabled by the associated Interrupt mask IE and the GIE bit is not set, it can wake up the controller from the sleep if interrupt occurs, but the Interrupt Service Routine located in the interrupt vector will not be executed and the code of the program will continue execution. The interrupt flag will set when its associate event occurs regardless of whether or not the GIE bit is set.

After switch is released, the RBIF flag is cleared and the device again is put to the sleep mode at the beginning of the loop..

```
    btfss  we          ; Is the switch released?

    goto $-1

    bcf intcon, rbif   : clear rbif flag

    goto loop   ;
```

Routine *del* causes 0.1 s delay calculated from the formula:

$$t = 11 \times 4 \times Tcm$$

11 decrements 256 times

4 machine cycles

Tcm = 108 us

Test 9. Working with debugger. Turn the LED on for the calculated period of time.

We will use MPLAB SIM debugging tool in the MPLAB IDE environment.

```
; ************************************************
; T1 Blinking the LED, connected to bit1 of PORT B,
;    with delay equal to 0,75 s.
;    Internal clock frequency 37 kHz, Tcm = 108 µs
; ************************************************

    list p=16f628, r=hex       ; declare processor,
                               ; specifying the radix
    #include p16f628.inc       ; include register label
                               ; definitions
    __config h'3f10'           ; configuration
                               ; information
                               ; for selected processor
    errorlevel -302            ; turn off banking
                               ; message
#define wy       portb,1       ; the directive
                               ; substitutes a text
                               ; portb,1 with a string
#define lop      .9            ;defines value of
                               ; 9 decimal
                               ; for register loop
cblock    h'20'   ; constant block
    l1, l2
endc
    movlw h'07'         ;07 -> w
    movwf cmcon         ; w->cmcon, comparators off
    clrf  porta         ; clear PORTA output latches
    clrf  portb         ;initializes PORTB
```

B. Borowik, *Interfacing PIC Microcontrollers to Peripherial Devices*, Intelligent Systems, Control and Automation: Science and Engineering 49, DOI 10.1007/978-94-007-1119-8_9,

```
    bsf   status, rp0 ;bank 1
    bcf   pcon, oscf        ;internal gen.32kHz,
                            ; Tcm=108µs
    clrf  trisa       ;PORTA for output
    clrf  trisb       ;PORTB for output

    bcf   status, rp0             ;bank 0
loop                        ; main loop
    bsf   wy          ; turn the LED on
    call  delay       ; call delay routine
    bcf   wy          ; turn off LED on RB1
    call  delay       ; call delay routine
    goto loop         ; repeat main loop

delay           ;delay 0,75 s (9*256*3*108 µs=0.746s)
    movlw lop         ; lop ->w
    movwf l2          ; initialize l2
    clrf  l1          ; initialize l1

del decfsz      l1, f       ; decrement l1
    goto  del         ; if not 0, keep decrementing
                      ; l1
    decfsz      l2, f ; decrement l2
    goto        del   ; if not 0, decrement l1 again
    return            ; return to main routine
end
; **************************************************
```

We select *Debugger>Select Tool* pull down menu and check MPLAB SIM. Additional menu items will appear in the Debugger menu.

The MPLAB SIM simulator is integrated into MPLAB IDE integrated development environment.

MPLAB SIM allows us to:

- Modify object code and immediately re-execute it
- Trace the execution of the object code

The simulator is a software model, and not actual device hardware.
When MPLAB SIM is simulating running in real time, instructions are executing as quickly as the PC's CPU will allow. This is usually slower than the actual device would run at its rated clock speed.
The speed at which the simulator runs depends on the speed of our computer and how many other tasks we have running in the background.
Once we have chosen a debug tool, we will see changes in the following on the IDE:

1. The status bar on the bottom of the MPLAB IDE window should change to "MPLAB SIM".
2. Additional menu items should now appear in the Debugger menu.
3. Additional toolbar icons should appear in the Debug Tool Bar. After positioning the mouse cursor over a toolbar button, a brief description of the button's function can be seen.
4. An MPLAB SIM tab is added to the Output window.

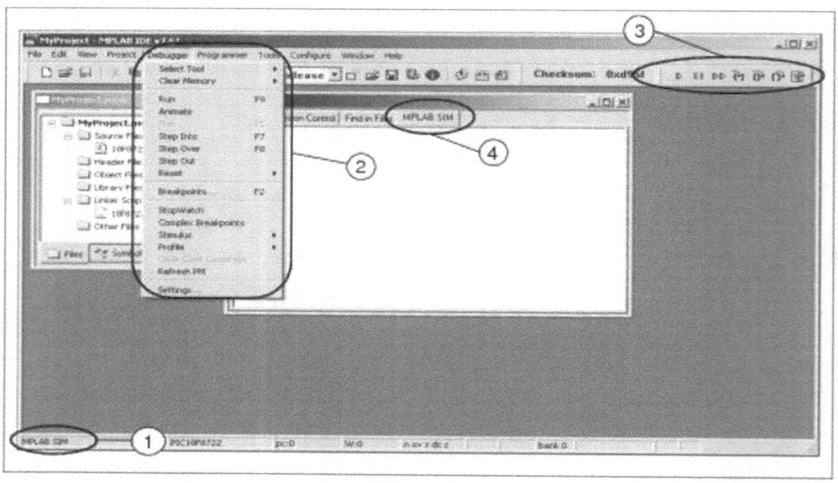

Fig. 12: MPLAB IDE Desktop with MPLAB SIM as Debugger

In the Debug Tool Bar there are tool bar icons:

Table . Table 2. Debug short cut icons.

Debugger Menu	Toolbar Buttons	Hot Key
Run		F9
Halt		F5
Animate		
Step Into		F7
Step Over		F8
Step Out		
Reset		F6

After positioning the mouse cursor over a toolbar button we can see a brief description of the button's function.

Standard debug windows allow to view the actual code, as well as the contents of program or data memory. The watch window displays the values of registers that are specified.

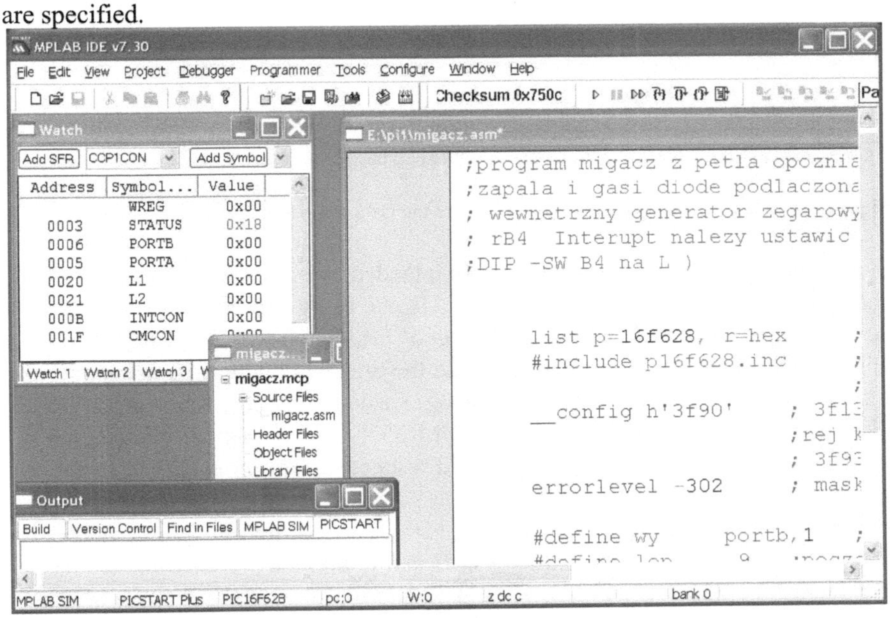

Fig. 13. Project migacz.mcp after selecting the built-in simulator MPLAB SIM.

Next we select Debugger -> Reset -> Processor Reset and a green arrow shows where the program will begin (see Fig. 14). In our project it is the line:

```
movlw h'07'
```

which is loading working register with the value of 7.

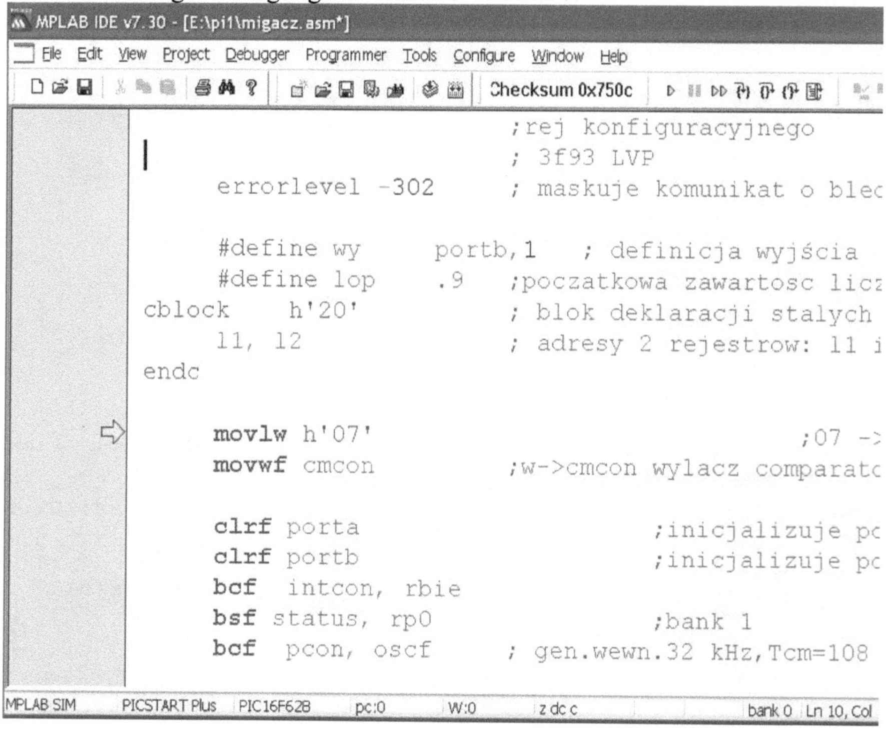

Fig. 14. After reseting a green arrow shows where the program will begin

In order to see if the code is operating as intended, we can watch the values being changed in the program. Selecting *View>Watch* brings up an empty Watch window. There are two pull downs on the top of the Watch window. The one on the left labeled "Add SFR" can be used to add the Special Function Registers into the watch. We select from the list following Special Function Registers: WREG, STATUS, PORTB, PORTA, INTCON, CMCON and sequentially after each selection we have to click **Add SFR** to add respective register each of them to the window.

The pull down on the right, allows symbols to be added from the program. We use this pull down to add two our variables: l1 and l2 into the Watch window. We select them from the list and then click **Add Symbol** to add them to the window. The Watch window should now show the address, value and name of the selected registers.

To single step through the application program, we select *Debugger>Step Into* or click the equivalent toolbar icon. This will execute the currently indicated line of code and move the arrow to the next line of code to be executed. After repeating twice *Step Into* the value of 7 will be moved from *WREG* register into SFR *CMCON.*

The value of CMCON register in the Watch window is in red color because it was changed by the previous debug operation, values that were not changed are black.

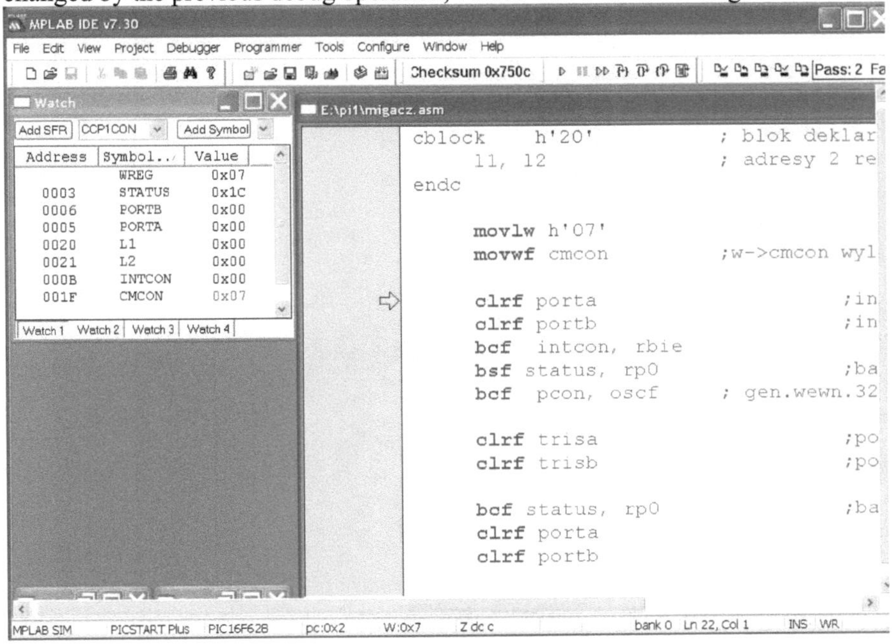

Fig. 15 CMCON register is in red and has the value of 0x07. Highlighted is PORTB register.

We can continue single stepping through the code. Next instructions are: clearing registers PORTA and PORTB and clearing RBIE bit of the INTCON register. Clearing the RBIE mask disables interrupt-on-change feature of PORTB. Then we will set RP0 bit of STATUS register for selecting data memory bank 1 as is shown in figure 16.

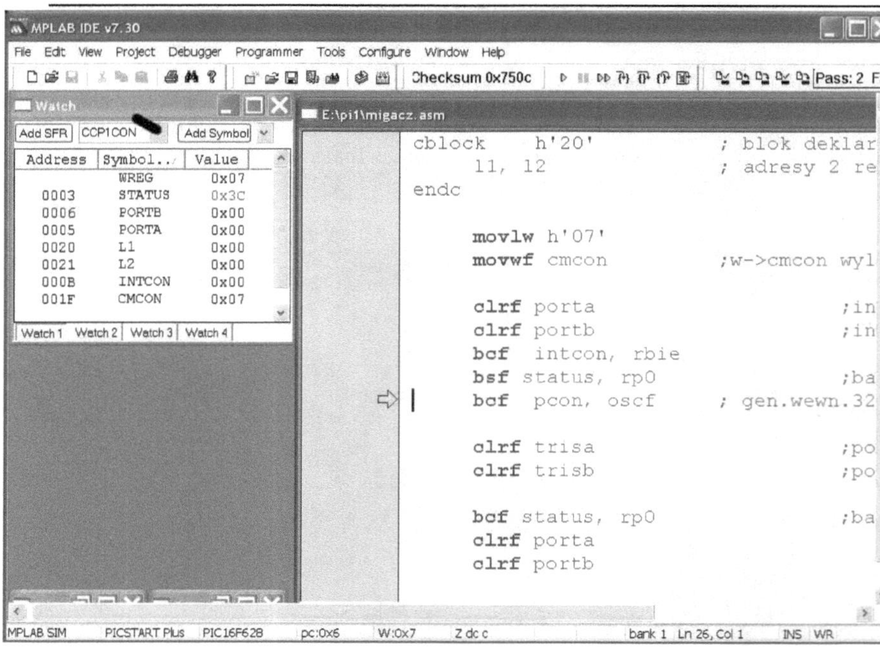

Fig. 16. STATUS register has the value of 0x3C in red

The STATUS register has the value of 0x3C, binary b'0011 1100'. Operation:

```
    bsf status, rp0    b
```

caused the bit RP0 (#5) to be set.
Steping into executes next instruction:

```
    bcf   pcon, oscf  ;internal gen.32kHz,Tcm=108µs
```

The direction registers TRISA ad TRISB are cleared, configuring all pins of PORTA ad PORTB as output.
After clearing PORTA and PORTB program enters the main loop.

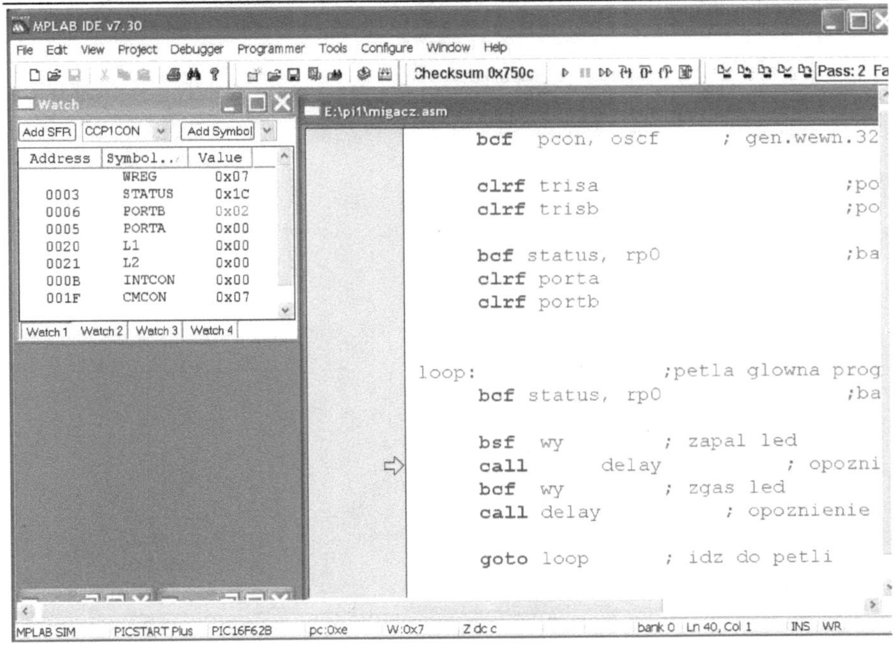

Fig. 17. Main loop

Two instructions:

```
        bsf   wy            ; turn the LED on

        bcf   wy            ; turn the LED off
```

set and clear bit 1 of PORTB. To that bit is attached LED diode. After setting bit PORTB,1 the value of PORTB register equals to 0x02, as is indicated in the Watch window.

Next instruction is :

```
        call  delay         ; call delay routine
```

After steping into we enter to the routine *delay*.

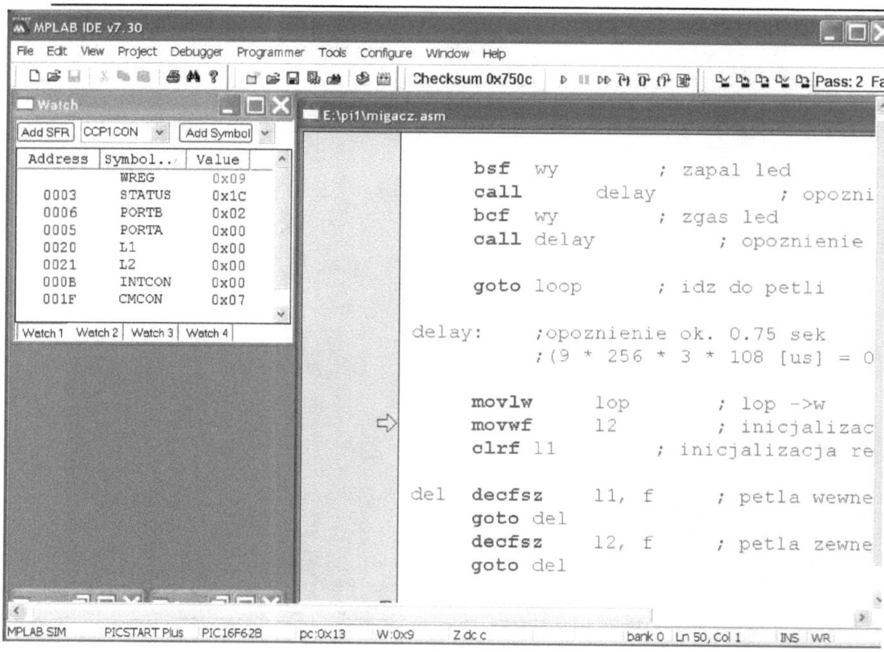

Fig. 18. delay procedure

Two user registers l1 and l2 are declared. They work as counters in the nested loop *del*.

```
#define lop     .9    ;defines value of 9 decimal
                      ; for variable lop
cblock    h'20'    ; constant block
     l1, l2
endc
```

General purpose static RAM registers can be declared in data memory space starting with address h20.
Register *l1* is first cleared. Then in the inner loop *(del)* it is decremented and tested if equal 0. After first decrementing from value 0 it receives value FF (255). Total count of decrements is 256. It alows for the delay of:

256 x 3 x 108 us = 82 944 us

256 – number of decrements

3 – number of machine cycles:

```
del     decfsz  l1, f     ; 1 machine cycle
        goto del ; 2 machine cycles
```

T_{CM} = 108 us

In the outer loop the counter *l2* is decremented starting with value of 9 to 0. Total delay of the nested loop would be about:

9 x 256 x 3 x 108 us = ca. 750 ms.

Let us watch next screens:

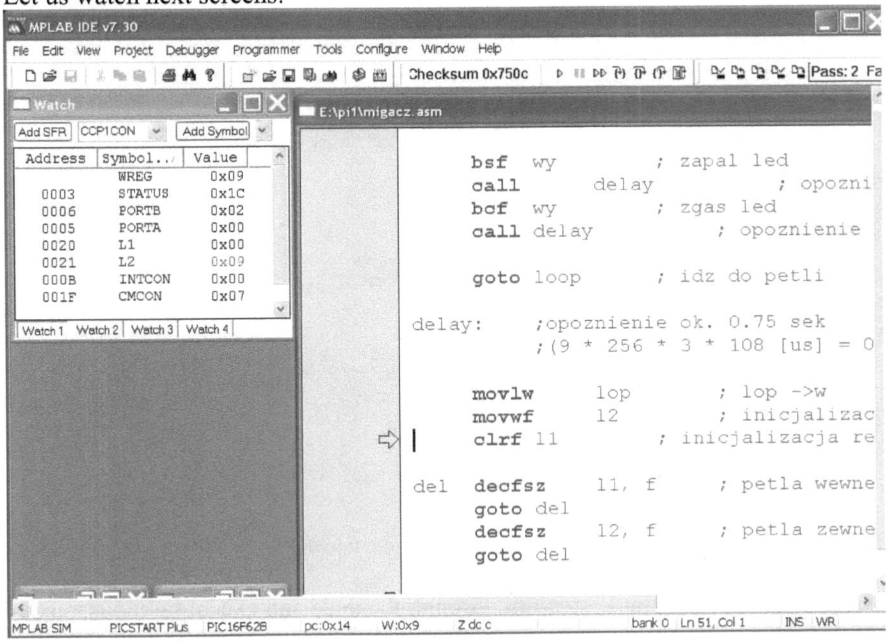

Fig. 19. Starting values of l1 and l2 in the window WATCH.

Starting values of l1 i l2 are 0x00 and 0x09 and such values are in the WATCH window.

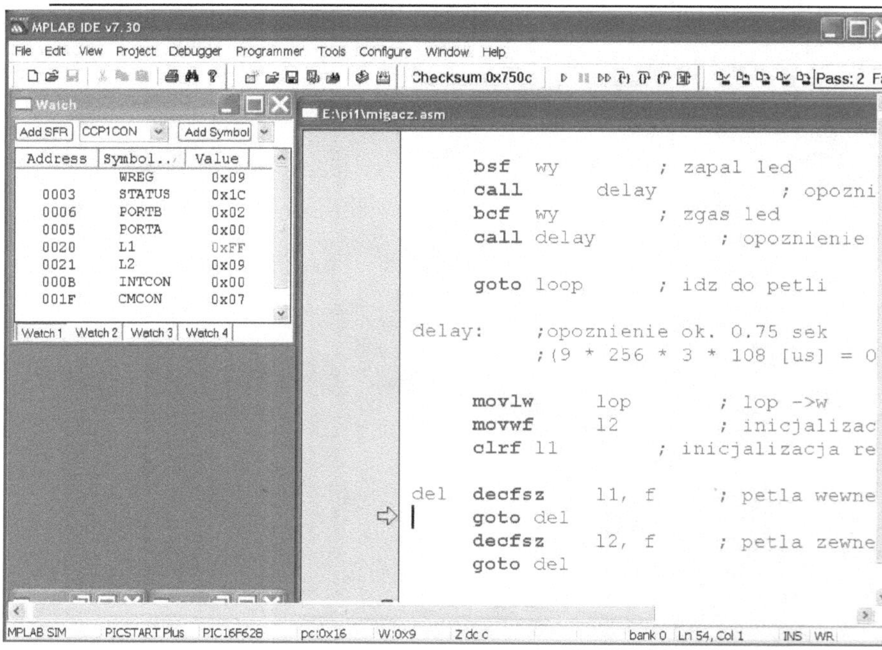

Fig. 20. After first decrementing *l1* changes its value from 0 to 0xFF

Instead of continuing single stepping through the code and executing thousands of steps, it is better to set a breakpoint on the selected line and run the code. To set a breakpoint, we can put the cursor on the selected line and click the right mouse button. Then select Set Breakpoint from the context menu. A red "B" will show on the line. (One can also double click on a line to add a breakpoint.)
On the figure below Breakpoint is set at the line, where *l2* variable is decremented and tested., after inner loop has been executed 256 times.
We can enter Run mode by either clicking the **Run** button on the Debug toolbar, selecting *Debugger>Run* from the menu bar, or pressing <F9> on the keyboard. Some tools provide additional types of run, such as "Run to cursor" from the right-mouse menu.

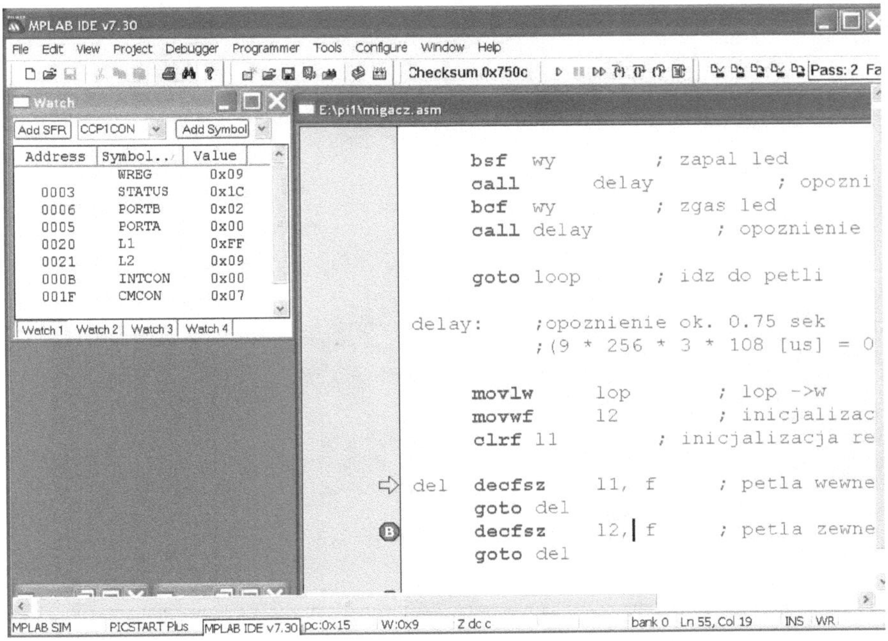

Fig. 21. Breakpoint set on the line *decfsz l2,f.*

After running application the program code is executed until a breakpoint is encountered. In the outer loop we decrement the variable *l2*. Next breakpoint should be set on the instruction *return*, in order to exit the subroutine *del* and return to main routine.

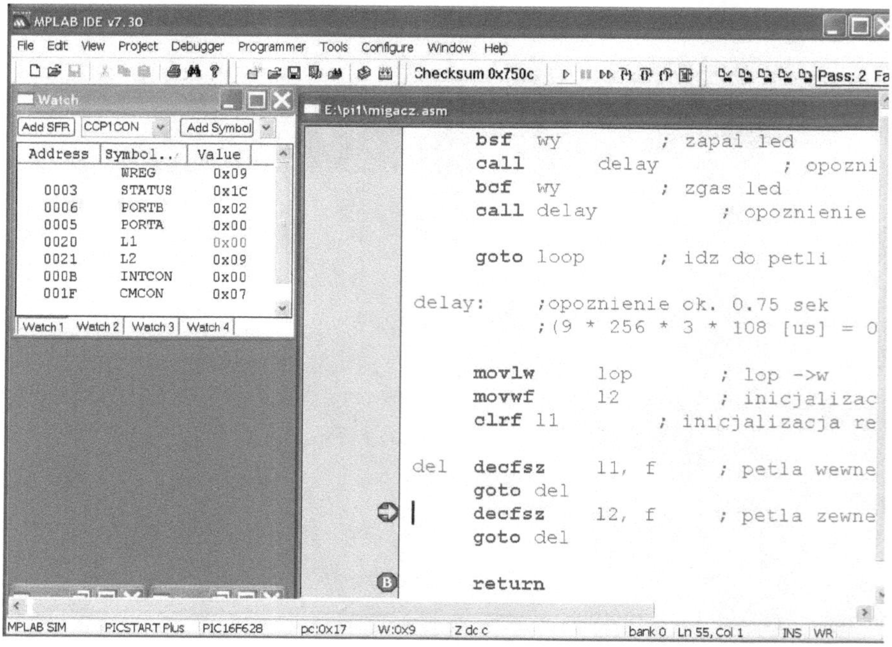

Fig. 22. Breakpoint set on the line *return* - returm to the main routine

Test 10. Driving a 7-Segment LED Display with PIC16F628 microcontroller

Below presented application was run on the prototype board Microcon4. The application can be used for testing all 8 lines connecting port B with the display and the display itself. 16 hexadecimal digits 0 – F are shown sequentialy one after another with time span 0.9s each.

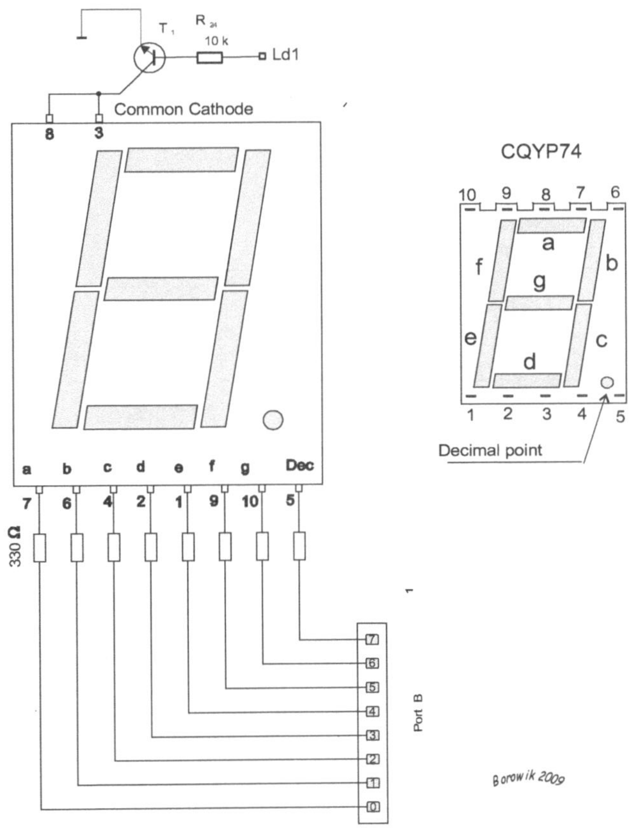

Fig. 23 Connecting a common cathode current switch T1 to 7-segment display

B. Borowik, *Interfacing PIC Microcontrollers to Peripherial Devices*, Intelligent Systems, Control and Automation: Science and Engineering 49, DOI 10.1007/978-94-007-1119-8_10,

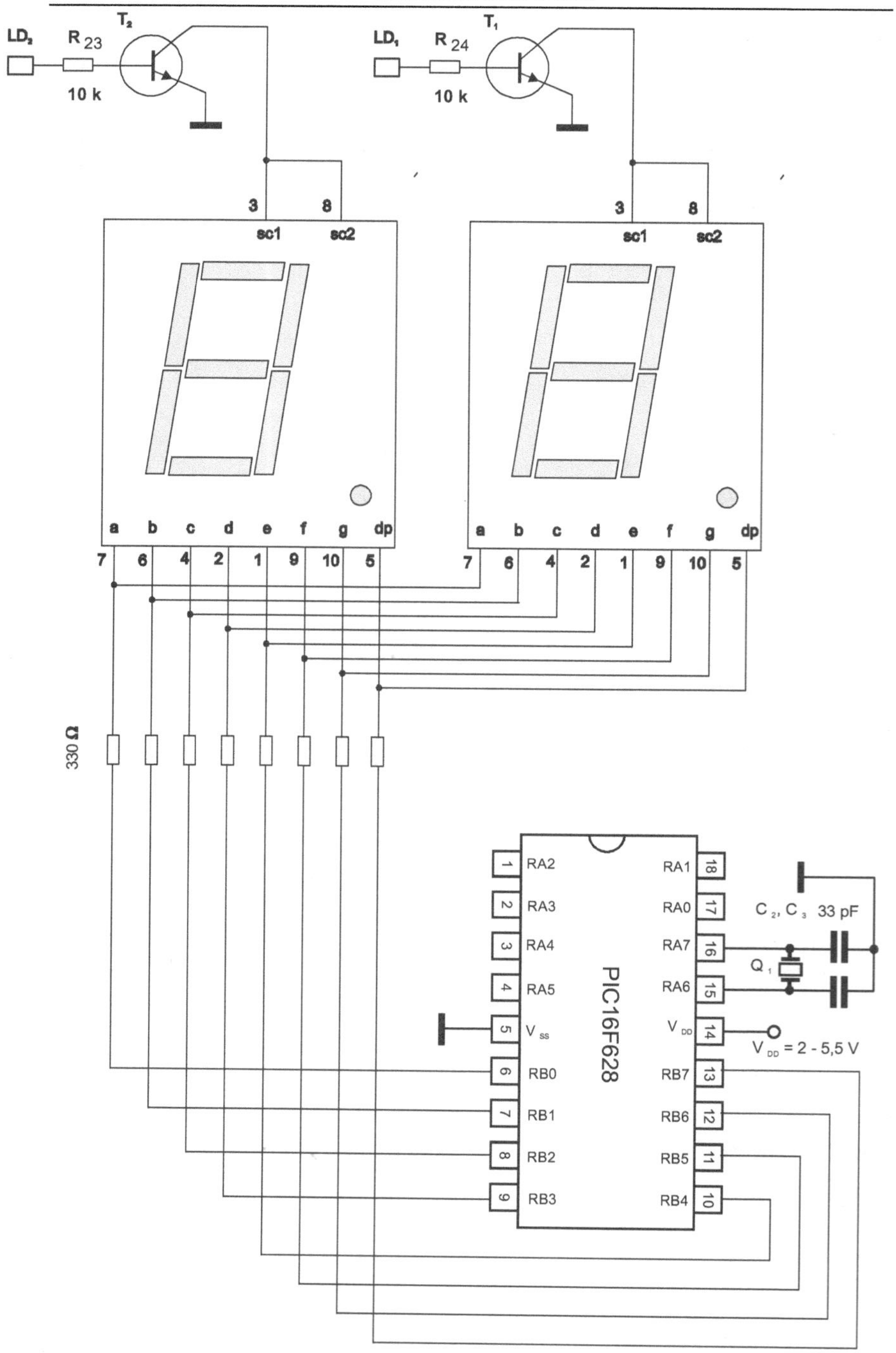

Fig. 24 Connecting a microcontroller to 7-segment display

The circuit is very simple. Port A, bits 0-4 are connected to the digit select lines directly. Port B, bits 0-7 are connected to the segment select lines through 330 Ω resistors. A 4MHz crystal resonator is connected to the OSC1/CLKIN and OSC2/CLKOUT pins of the PIC, with the center lead connected to Gnd. Vss and Vdd are connected to Gnd and +5V, respectively.

```
; program 10

      list p=16f628, r=hex      ; declare processor,
                                ; specyfying the radix
      #include p16f628.inc      ; include register label
                                ; definitions
      __config h'3f10'          ; configuration
                                ; information
                                ; for selected processor
                    ; 3f10 HVP (High Voltage Programming)
                    ; 3f90 LVP (Low Voltage Programming)
#define czas    .8
l1   equ   h'20'
l2   equ   h'21'

d0   equ   b'00111111' ; digit 0
d1   equ   b'00000110'
d2   equ   b'01011011'
d3   equ   b'01001111'
d4   equ   b'01100110'
d5   equ   b'01101101'
d6   equ   b'01111101'
d7   equ   b'00000111'
d8   equ   b'01111111'
d9   equ   b'01101111'
diga       equ   b'01110111' ; digit A
digb       equ   b'01111100'
digc       equ   b'01011000'
```

```
digd        equ    b'01011110'
dige        equ    b'01111001'
digf        equ    b'01110001'
dig7        equ    b'10000000' ; decimal point

      movlw h'07'              ;07 -> w
      movwf      cmcon         ; w->cmcon,
                               ; comparators off
      bsf status, rp0                ;bank 1
    bcf   pcon, oscf           ;internal gen.32 kHz,
                               ; Tcm=108µs

    clrf  trisa         ;PORTA for output
    clrf  trisb         ;PORTB for output

    bcf status, rp0                  ;bank 0

    clrf  porta         ; clear PORTA output latches
    clrf  portb         ;initializes PORTB

    bcf porta, 6                     ;zgas  LED
    bcf porta, 2                     ;LED 2 off
    bsf porta, 0               ; LED 0 on

       bsf porta, 7                  ; LED 7 on
       bcf porta, 5                  ; LED 5 off
       bcf porta, 3                  ; LED 3 off
       bcf porta, 1                  ; LED 1 off

    nop

    movlw d0
```

```
movwf        portb
  call przerwa

  movlw d1
  movwf             portb
  call przerwa

  movlw d2
  movwf             portb
  call przerwa

  movlw d3
  movwf             portb
  call przerwa

  movlw d4
  movwf             portb
  call przerwa

  movlw d5
  movwf             portb
  call przerwa

  movlw d6
  movwf             portb
  call przerwa

  movlw d7
  movwf             portb
  call przerwa

  movlw d8
```

```
movwf          portb
call przerwa

movlw d9
movwf          portb
call przerwa

movlw diga
movwf          portb
call przerwa

movlw digb
movwf          portb
call przerwa

movlw digc
movwf          portb
call przerwa

movlw digd
movwf          portb
call przerwa

movlw dige
movwf          portb
call przerwa

movlw digf
movwf          portb
call przerwa

movlw dig7
```

```
          movwf              portb
          call przerwa


        goto $               ; go to self
                             ; loop here forever
    przerwa
        movlw czas
        movwf        l1
        clrf  l2
    spr         nop
        decfsz       l2,f
        goto spr
        decfsz       l1,f
        goto  spr
    return


    end
```

Program description:
The directive:

```
#define czas    .8
```

defines value of 8 decimal for register variable *czas.*
In the MPASM assembler (the assembler) that provides a platform for developing assembly language code for Microchip's PICmicro microcontroller (MCU) families, a decimal integer is `d' or `D' followed by one or more decimal digits `0123456789' in single quotes, or, a decimal integer is `.' followed by one or more decimal digits `0123456789'.
The directives:

```
l1   equu   h'20'
l2   equu   h'21'
```

are used to assign a variable names *l1* and *l2* to an address locations in RAM respectively h'20' and h’21’.
Routine *przerwa* contains nested loop. Two general purpose registers l2 and l1 are decremented. l2 is first cleared and then is decremented in the inner loop with instruction decfsz (decrement f, skip if zero). The loop is executed 256 times. It takes time of 110.592ms (256 x 4 x 108us).

4 accounts for 4 machine cycles:

```
spr   nop              ;        1 machine cycle
      decfsz   12,f    ;        1 machine cycle
      goto spr ;            2 machine cycles
```

For internal gen.32 kHz,Tcm=108μs
L1 user register is decremented 8 times.
Loop requires to execute:

t = 108 * 4 * 8 * 256 = 884736 us = 885 ms = about 0.9 s

Nop instruction adds up to the number of machine cycles in one iteration.
The segments in a 7-segment display are arranged to form a single digit from 0 to F as shown in the figure:

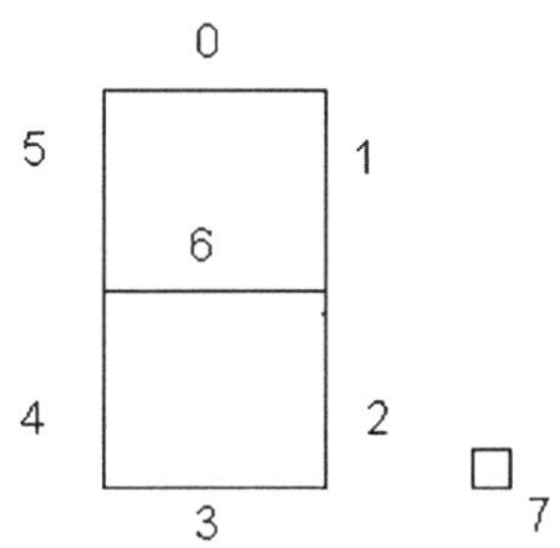

Fig. 29 Segments of the 7-segment Display.

Segments 0 to 7 are marked respectively with non-capital letters: a, b, c, d, e, f, g and dp, where dp is the decimal point. The 8 display segments lines (0 to 7) are connected to the 8 PortB lines. PORTB is made output by clearing TRISB register.
In order to reduce the number of pins in the device, a clever arrangement of common cathode is employed. Since there are 8 segments (7 plus the decimal point), there are 8 anode connections. Onto the PORTB are put respective values, representing particular digits from 0 to F and at the last: the decimal point. Those values were defined at the beginning of the program.
Value 1 means the respective LED segment is turned on. To display 8 we need to put a binary b'01111111' on PortB. This will turn on all the LEDs except the decimal point. The number 0 is simply an 8 with the middle segment off, b'00111111'. The decimal point has its value: b'10000000'.

Test 11. Driving a 7-Segment LED Display with PIC16F628 microcontroller (cont.)

Our board is containing a dual 7 segment LED display. The display is a common catode type, so to light a segment we need to take the relevant segment high and the common cathode must be connected to the 0V rail. In order to reduce the I/O count we're multiplexing the display digits.
Program prints two digits: '0 1' on two 7seg displays in multiplex mode. Displays are multiplexed with a frequency of 61 Hz:

T multipl. = 1 us * 64 (prescaler) * 256 (counter) = 16 384
1/T = appr. 61 Hz

Code listing:

```
list p=16f628, r=hex     ; declare processor,
                         ; specyfying the radix
#include p16f628.inc     ; include register label
                         ; definitions
__config h'3f10'         ; configuration
                         ; information for
                         ; selected processor
            ; 3f10 HVP (High Voltage Programming)
            ; 3f90 LVP (Low Voltage Programming)
b_wys       equ   h'20' ; 2 bytes buffer b_wys
                         ; for storing digits
                         ; for each display
nr_wys      equ   h'22' ; 7-segment display
                         ; number (0 or 1)
w_temp      equ   h'23' ; auxiliary variable for
                         ; storing W register
                         ; content
st_temp     equ   h'24' ; auxiliary for storing
                         ; status register
temp_porta  equ   h'25' ; auxiliary for storing
                         ; porta register
```

B. Borowik, *Interfacing PIC Microcontrollers to Peripherial Devices*, Intelligent Systems, Control and Automation: Science and Engineering 49, DOI 10.1007/978-94-007-1119-8_11,

```
        org    0
        goto   start          ; beginning of the program
        org    4
_int          ; beginning of the interrupt routine
        ; store the contents of W and STATUS registers
        movwf         w_temp        ; w_temp <- w
        swapf         status, w
        movwf         st_temp       ; st_temp <- w
        bcf           status, RP0        ; bank0
        bcf           intcon, T0IF       ; clear flag
                            ; overflow TMR0 to be able
                            ; to react to the next
                            ; interrupt
        movlw         0x01
        movwf         tmr0               ; initialize TMR0
                            ; to have the next interrupt
                            ; in approximately 16 ms

        ; changing the display:       1 -> 0
                                ;     0 -> 1
        incf          nr_wys, f     ; incrementing
        movf          nr_wys, w     ; w <- nr_wys
        xorlw         0x02          ; equal to 2 ?
                                    ; If so, set Z flag
        btfsc         status, Z     ;
        clrf          nr_wys        ;

                      ; clearing RA0, RA1
        movf          temp_porta, w
        andlw         0xfc          ; b'1111 1100'
        movwf         temp_porta
```

```
    movf        nr_wys, w
    call        _poz
    iorwf       temp_porta, f       ; w=w or temp_porta
    movlw       b_wys               ; beginning of the
                                    ; display buffer
    addwf       nr_wys, w           ; W=b_wys + nr_wys
    movwf       fsr
    movf        indf, w       ; read the digit from RAM
    movwf       portb         ; put the digit into the
                              ; display

    movf        temp_porta, w
    movwf       porta

    ; restoring registers

    swapf       st_temp, w
    movwf       status
    swapf       w_temp, f
    swapf       w_temp, w
    retfie

_poz        addwf       pcl,f
    dt          1,2           ; retlw 1
                              ; retlw 2

start                               ; the main program
    bcf         status, RP0         ; bank 0
    movlw       h'07'
    movwf       cmcon
    clrf        porta
    clrf        portb
```

```
bsf         status, RP0 ; bank 1
bsf         pcon, oscf         ; 4 MHz
bcf         option_reg, t0cs   ; Internal
                  ; instruction cycle clock
                  ; fxtal/4
bcf         option_reg, PSA    ; prescaler is
                  ; assigned to the Timer0 module
bsf         option_reg, PS0    ; prescaler rate
                               ; select bits: 1/64
bcf         option_reg, PS1    ; PS0, PS1, PS2=101
bsf         option_reg, PS2
movlw       h'00'
movwf       trisb        ; PORTB for output
clrf        trisa        ; PORTA for output
bcf         status, RP0        ; BANK 0
clrf        nr_wys       ; display # 0

call   _int
movlw       0ffh
movwf       porta        ;initialize
movwf       portb
movwf       temp_porta

clrf        intcon       ; disabling all
                         ; interrupts
movlw       001h         ; starting value of TMR0,
                         ; thus interrupt occurs
                         ; every 255 x 64 =16 ms
movwf       tmr0

bsf   intcon, T0IE       ; enables the TMR0
                         ; interrupt
```

```
    bsf   intcon, GIE          ; enables all interrupts

    call _wyswietlanie         ; reading digits '0'
                               ; or '1' to the buffer

_et nop                        ; main loop
    goto  _et

_wyswietlanie
    movlw       b_wys          ; display buffer
    movwf       fsr
    movlw       3fh            ; store digit '0' (3fh)
    movwf       indf           ; in the first byte

    incf        fsr,f
    movlw       06h            ; store digit '1'
    movwf       indf           ; in the second byte

    return
    end
```

Program description:

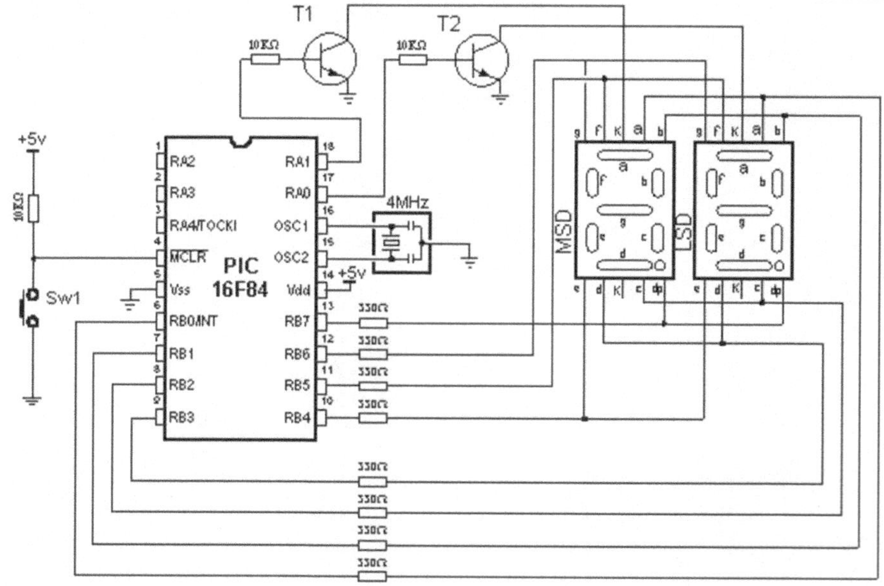

Fig. .26 Connecting a microcontroller to 7-segment displays in multiplex mode

Each display is turned on at a such rate, that persistence of vision of our eye thinks the display is turned on the whole time. As each display is turned on, the appropriate information must be delivered to it so that it will give the correct reading. Therefore, the program has to ensure the proper timing, else the unpleasant blinking of display will occur.

Displaying digits is carried out in multiplex mode which means that the microcontroller alternately prints on first and on the second display. TMR0 interrupt serves for generating a time period, so that the program enters the interrupt routine every 16 ms and performs multiplexing. In the interrupt routine, first step is deciding which segment display should be turned on. In case that the first display was previously on, it should be turned off, set the mask for printing the digit on next 7seg display which lasts 16 ms, i.e. until the next interrupt.

Operating the device is fairly straight forward. By selectively connecting a digit-select line (one of the 2 common cathodes) to ground and applying a positive voltage to any of the segment-select lines (common anode), it is possible to illuminate any segment of any digit. Port B on the PIC16F628 is 8 bits wide: one bit for each segment.

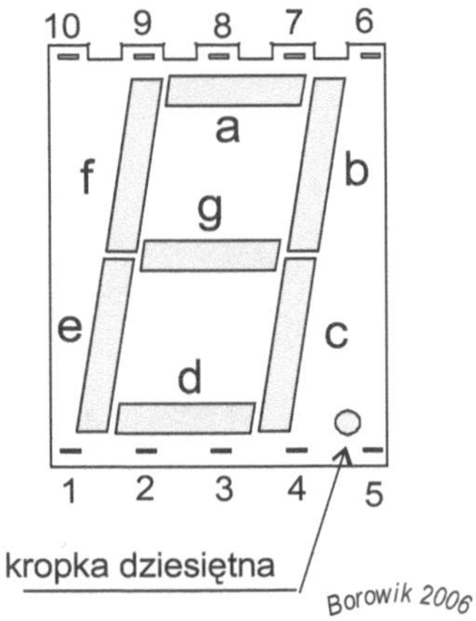

Fig. 27 7 segment display

For displaying digit ‘0’ we have to turn on the following segments of the display: a, b, c, d, e, f. Into port B we have to put the value binary: b’00111111’, hexadecimaly: 0x3f.

For displaying digit ‘1’ we have to turn on the segments b and c. Into port B we have to put the value binary: b’00000110’, hexadecimaly: 0x06.

Both values in the routine `_wyswietlanie` we put into RAM memory to the variable `b_wys`, denoting the display buffer.

```
        b_wys           equ    h'20' ; 2 bytes buffer b_wys
                                     ; for storing digits
                                     ; for each display

    _wyswietlanie
        movlw           b_wys        ; display buffer
        movwf           fsr
        movlw           3fh          ; store digit '0' (3fh)
        movwf           indf         ; in the first byte

        incf            fsr,f
        movlw           06h          ; store digit '1'
        movwf           indf         ; in the second byte
        return
```

Let us see the code snipett. Two displays are numbered: 0 and 1.

```
; changing the display:         1 -> 0
                        ;       0 -> 1
incf        nr_wys, f   ; incrementing
movf        nr_wys, w   ; w <- nr_wys
xorlw       0x02        ; equal to 2 ?
                        ;If so, set Z flag
btfsc       status, Z   ;
clrf        nr_wys      ;
```

After incrementing display # 0 becomes # 1. That is OK.
Incrementing display number 1 gives #2. We detect this by xoring nr_wys (display #) with the value of 2 (0x02). Then we clear it, so display # 1 becomes # 0.

```
            ; clearing RA0, RA1
movf        temp_porta, w
andlw       0xfc        ; b'1111 1100'
movwf       temp_porta
```

We put to ground common cathodes of both displays, and save this setting to auxiliary variable temp_porta.. Now we have to send the appropriate digit pattern to port B. Both digit patterns are stored in RAM memory in the INDF register. The first digit: 0 (the pattern is 3fh or b'0011 1111') has the memory address h20. The next digit: 1 (the pattern 06h or b'0000 0110') occupies the next memory location.

```
    movf        nr_wys, w
    call        _poz
```

In above two lines we load the display # into W register. Then we call subroutine *_poz*. If the display # was 0, we receive in W on return value of 1. If the display # was 1, we receive in W on return value of 2.

```
_poz    addwf       pcl,f
    dt          1,2         ; retlw 1
                            ; retlw 2
```

This value in W register is then ored with temp_porta register in order to turn off one of two 7-segment displays. If the W register was 1, the RA0 receives value of 1, turning the attached to it 7-segment display off. If the W register was 2, the RA1 receives value of 1, turning the attached to it 7-segment display off.

```
iorwf        temp_porta, f      ; w=w or temp_porta
```

Then the address in RAM memory of the digit pattern to be displayed is composed. To the address of the display buffer *b_wys*, which is 0x20 we add the display #, which is 0 or 1. Thus W now points to the digit pattern of '0' (0x3F) or to the digit pattern of '1' (0x06). Contents of W register is then moved to the File Select Register and appropriate digit pattern can be read from the *indf* register. Obtained digit pattern is next put into port B.

```
movlw        b_wys          ; beginning of the
                            ; display buffer
addwf        nr_wys, w      ; W = b_wys + nr_wys
movwf        fsr
movf         indf, w        ; read the digit from RAM
movwf        portb          ; put the digit into the
                            ; display
```

Of course, to activate 7-segment display, the content of the auxiliary variable *temp_porta* has to be sent to PORTA register.

```
movf         temp_porta, w
movwf        porta
```

Restoring the contents of W and STATUS registers is the last task in the interrupt service routine _int.

```
             ; restoring registers
swapf        st_temp, w
movwf        status
swapf        w_temp, f
swapf        w_temp, w
retfie
```

Test 12. Interfacing a PIC microcontroller to an LCD Hitachi Display.

We used an LCD 2x 16 HMC 16225 S-PY alphanumeric display module that is HD44780 compatible. Display has two lines, 16 characters each.
We work in 4-bit data transfer mode and we write strings to the upper and lower LCD lines.

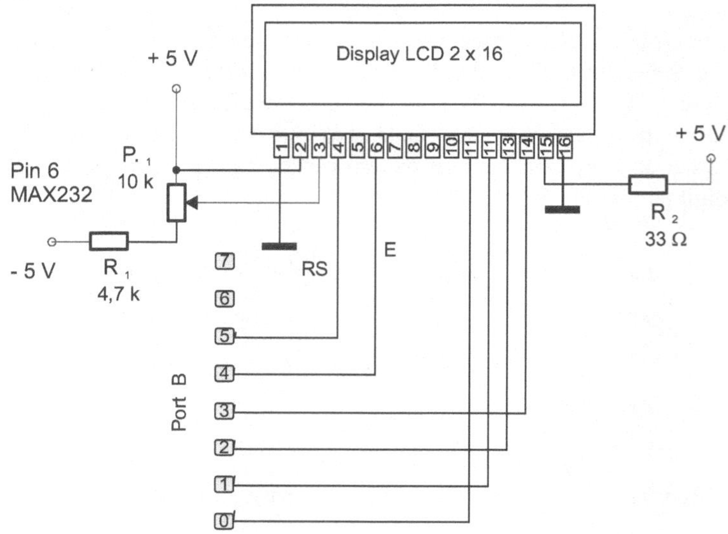

Fig. 28 Connecting an LCD display to a microcontroller

On the testing board used, the port B of the PIC microcontroller is dedicated for interfacing with the LCD module. It is connected to SW4 dip switch. Switches settings are as follows:

SW1 and SW2 set to High.
SW3 off
SW4 on.

We clear TRISB, so that PORTB is made output.
Low nibble of the port B: lines PORTB <3:0> communicate with DB7(pin 14) : DB4 (pin 11) of the LCD Display. LCD control signals E (Enable) and RS (Register Select) are connected to the RB4 and RB5 microcontroller PORTB lines respectively.
In 4-bit mode, two transfers per character / command are required. Half of the byte is sent in one operation, upper nibble first.
LCD's (drivers) are slow devices when compared to microcontrollers. Care must be taken from having communication occur too quickly. The software will need to control communicaton speed and timing to ensure the slow LCD and fast microcontroller can stay synchronized.

B. Borowik, *Interfacing PIC Microcontrollers to Peripherial Devices*, Intelligent Systems, Control and Automation: Science and Engineering 49, DOI 10.1007/978-94-007-1119-8_12,

When we start to communicate with the LCD module, we follow a standard LCD initialization sequence as recommended by the manufacturer. The initialization sequence must be timed precisely (see the HD44780 instruction set for details) and cannot be initiated before at least 30 ms have been granted to the LCD module to proceed with its own internal initialization (power on reset) sequence.

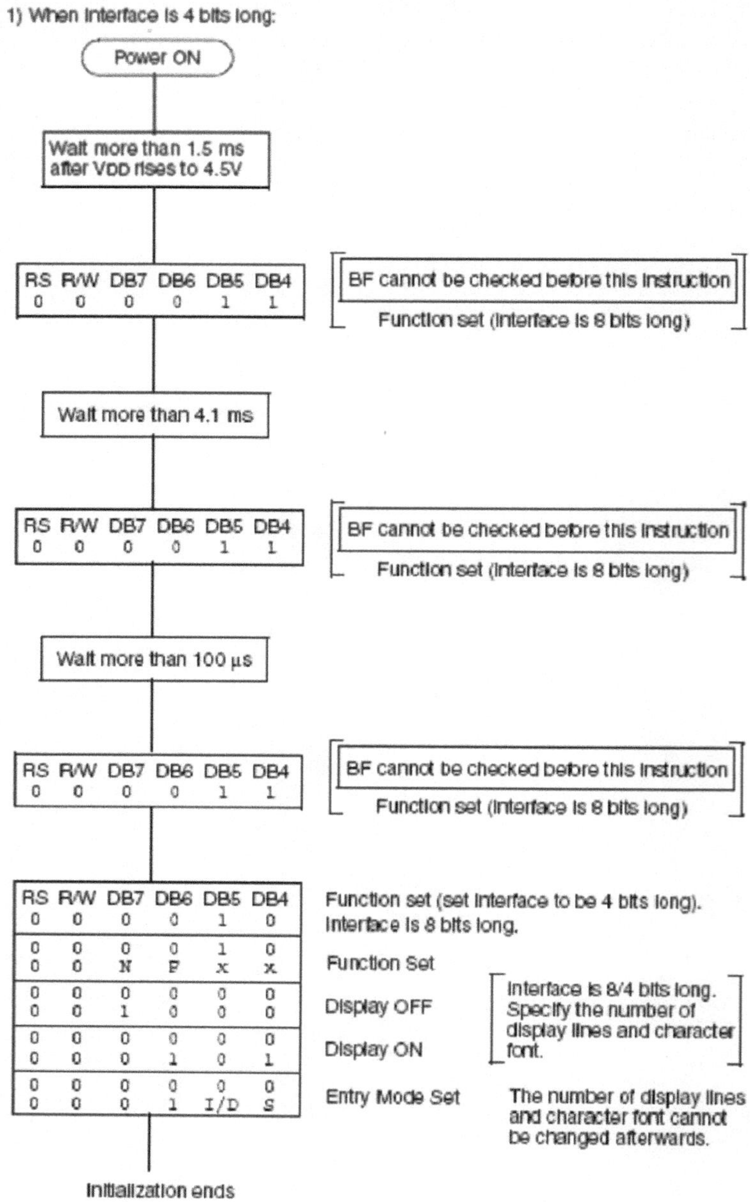

Fig. 29 Initialization flow for LCD module:

When the module powers up, the default data transfer mode is 8-bit. The initialization sequence only requires commands that are 4-bit in length. The last initialization command needs to specify the data transfer width (4-or 8-bit). Then a delay of 4.6 ms must be executed before the LCD module can be initialized.

Table 2 The HD44780 instruction set:

Instruction	Code										Description	Execution Time
	RS	R/W	DB7	DB6	DB5	DB4	DB3	DB2	DB1	DB0		
Clear display	0	0	0	0	0	0	0	0	0	1	Clears display and returns cursor to the home position (address 0).	1.64 mS
Cursor home	0	0	0	0	0	0	0	0	1	*	Returns cursor to home position (address 0). Also returns display being shifted to the original position. DDRAM contents remain unchanged.	1.64 mS
Entry mode set	0	0	0	0	0	0	0	1	I/D	S	Sets cursor move direction (I/D), specifies to shift the display (S). These operations are performed during data read/write.	40 uS
Display on/off control	0	0	0	0	0	0	1	D	C	B	Sets on/off of all display (D), cursor on/off (C), and blink of cursor position character (B).	
Cursor/ display shift	0	0	0	0	0	1	S/C	R/L	*	*	Sets cursor move or display shift (S/C), shift direction (R/L). DDRAM contents remain unchanged.	40 uS
Function set	0	0	0	0	1	DL	N	F	*	*	Sets interface data length (DL), number of display lines (N), and character font (F).	
Set CGRAM address	0	0	0	1	CGRAM address						Sets the CGRAM address. CGRAM data is sent and received after this setting.	40 uS
Set DDRAM address	0	0	1	DDRAM address							Sets the DDRAM address. DDRAM data is sent and received after this setting	40 uS
Read busy flag and address counter	0	1	BF	CGRAM/DDRAM address							Reads busy flag (BF), indicating internal operation is being performed, and reads CGRAM or DDRAM address counter contents (depending on previous instruction).	0 uS
Write to CGRAM or DDRAM	1	0	write data								Writes data to CGRAM or DDRAM.	40 uS
Read from CGRAM or DDRAM	1	1	read data								Reads data from CGRAM or DDRAM.	40 uS

Table 3 HD44780 command bits:

Bit Name	Setting/Status	
I/D	0 = Decrement cursor position	1 = Increment cursor position
S	0 = No display shift	1 = Display shift
D	0 = Display off	1 = Display on
C	0 = Cursor off	1 = Cursor on
B	0 = Cursor blink off	1 = Cursor blink on
S/C	0 = Move cursor	1 = Shift display
R/L	0 = Shift left	1 = Shift right
DL	0 = 4-bit interface	1 = 8-bit interface
N	0 = 1/8 or 1/11 Duty (1 line)	1 = 1/16 Duty (2 lines)
F	0 = 5 × 7 dots	1 = 5 × 10 dots
BF	0 = Can accept instruction	1 = Internal operation in progress

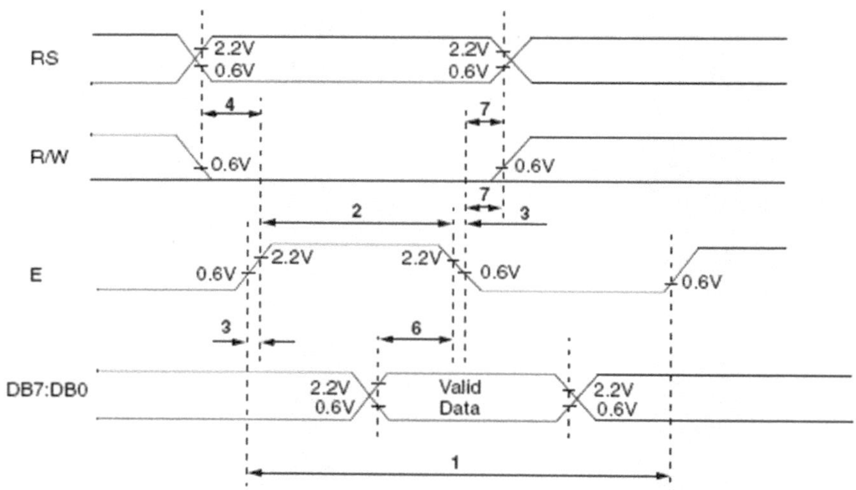

Fig. 29 Data write interface timing

The LCD controller needs 40 to 120 microseconds (uS) for writing and reading. Other operations can take up to 5 mS. During that time, the microcontroller can not access the LCD, so a program needs to know when the LCD is busy. We can solve this in two ways.

One way is to check the LCD status register BUSY bit (BF flag) found on data line D7 (LCD pin 14). When it is low, data is written to LCD, the LCD is ready for next operation. When it is high, operation is in progress and we have to wait. On the testing board MICROCON4, R/W line is grounded and always in low therefore reaing from the LCD module status-register is disabled.
The other way to ensure, LCD is not busy, is to introduce a delays in the program. The delays have to be long enough for the LCD to finish the operation in process. The practice shows, that following delays are needed for LCD operations:

50 ms

4,1 ms

100 μs

40 μs

Those are minimal values. Considering that the typical LCD module is an extremely slow device, we will better select the most generous timing, adding the maximum number of wait states, allowed at each phase of a read or write sequence:

4 x T_{MC} wait for data set up before read/write

15 x T_{MC} wait between R/W and enable

4 x T_{MC} wait data set up after enable

In our experiment the following delay procedures were made:

p50 (app. 50 ms)
p5 (app. 5 ms)
p100 (app 100 μs)

When the LCD is initialized, it is ready to continue receiving data or instructions. If it receives a character, it will write it on the display and move the cursor one space to the right. The Cursor marks the next location where a character will be written. When we want to write a string of characters, first we need to set up the starting address, and then send one character at a time.

Program code lcd2.asm

```
;LCD2 text demo - 4 bit mode
    list p=16f628, r=hex        ; declare processor,
                                ; specyfying the radix
    #include p16f628.inc        ; include register label
                                ; definitions
    __config h'3f10'            ; configuration
                                ; information
```

```
                                     ; for selected processor
                          ; 3f10 HVP (High Voltage Programming)
                          ; 3f90 LVP (Low Voltage Programming)

      cblock      0x20
            TMP
            adres
            a1
            a2
      endc
#define E portb,4
#define RS      portb,5
      org 0
      bsf   status, rp0
      movlw 0x00
      movwf trisb
      bsf   pcon, oscf  ; 4 MHz
      bcf   status, rp0
      call initLCD
      movlw 80h   ; address of the first character,
                  ; first line
      call piszin
      call wys_strona

      movlw 0c0h  ; address of the first character,
                  ; second line
      call piszin
      call wys_adres
      goto  $

initLCD
      call p50
```

```
    movlw 0x13          ; E (RB4) = 1,
                        ; data to be sent (RB3:RB0)=3
                        ; w = 0001 0011
    movwf PORTB         ; w -> PORTB
    nop
    nop
    bcf E        ; the Enable strobe asserted low
    call p5
    bsf E        ; the Enable strobe asserted high
    nop
    nop
    bcf   E
    call p100
    bsf E
    nop
    nop
    bcf   E
    call p100
    movlw h'28' ; 4 bits mode, 2 lines, 5x7
    call piszin
    movlw .8            ; display off
    call piszin
    movlw .1            ; display clear
    call piszin
    movlw .6            ; entry mode
    call piszin
    movlw 0ch           ; display on
    call piszin
    return

;delay 100 us
P100:  ; wait t = 5 + 25*4 cycles
```

```
    movlw 0x01                  ;1 cycle
    movwf a1                    ;1 cycle
Out3:
    movlw 0x19                  ;1 cycle
    movwf 0x0E                  ;1 cycle
In3:
    decf 0x0E,1                 ;1 cycle
    btfss STATUS,Z              ;1 cycle (Z=0),
                                ;2 cycles (Z=1)
    goto In3                    ;2 cycles
    decf a1,1                   ;1 cycle
    btfss STATUS,Z              ; ;1 cycle (Z=0),
                                ;2 cycles (Z=1)
    goto Out3                   ;2 cycles
    return                      ;2 cycles
;----------------------------
; delay 2 ms
P2:  ; wait  t = 4 + 10 * (6 + 50 * 4)cycles
    movlw 0x0A                  ;1 cycle
    movwf a1                          ;1 cycle
Out2:
    movlw 0x32                  ;1 cycle
    movwf 0x0E                  ;1 cycle
In2:
    decf 0x0E,1                 ;1 cycle
    btfss STATUS,Z              ;1 cycle (Z=0),
                                ;2 cycles (Z=1)

    goto In2                    ;2 cycles
    decf a1,1                   ;1 cycle
    btfss STATUS,Z              ;1 cycle (Z=0),
                                ;2 cycles (Z=1)
```

```
   goto Out2                  ;2 cycles
   return                     ;2 cycles
;----------------------------

; delay 5 ms
P5:         ; wait t = 4 + 21 * (6 + 58 * 4)cycles
            ; = 4 + 21*238 = 4 + 4 998 us
   movlw 0x15                 ; 1 cycle
   movwf a1                   ; 1 cycle
Out1:
   movlw 0x3A                 ;1 cycle
   movwf 0x0E                 ;1 cycle
In1:
   decf 0x0E,1                ;1 cycle
   btfss STATUS,Z             ;1 cycle (Z=0),
                              ;2 cycles (Z=1)
   goto In1                   ;2 cycles
   decf a1,1                  ;1 cycle
   btfss STATUS,Z             ;1 cycle (Z=0),
                              ;2 cycles (Z=1)
   goto Out1                  ;2 cycles
   return                     ;2 cycles
;----------------------------
; delay 50
P50:      ; wait  T = 10 * 5000 cycles
   movlw 0x0A                 ;1 cycle
   movwf a2                   ;1 cycle
Jedziemy2:
   call P5                    ;2 cycles
   decf a2,1                  ;1 cycle
   btfss STATUS,Z         ;1 cycle (Z=0),
                              ;2 cycles (Z=1)
```

```
  goto Jedziemy2          ;2 cycles
  return                  ;2 cycles
;-----------------------------

piszd
    movwf TMP
    bcf E
    swapf TMP,f
    movlw 0c0h
    iorwf TMP,w
    andlw 0efh
    movwf portb          ; high nibble transfer
    bsf RS
    bsf E
    nop
    nop
    bcf E
    swapf tmp,w
    iorlw 0c0h
    andlw 0efh
    movwf portb          ; ; low nibble transfer
    bsf RS
    bsf E
    nop
    nop
    bcf   E
    call p100
    return
piszin                   ; RS = 0
    movwf TMP
    bcf E
    swapf TMP,f
```

```
    movlw 0c0h
    iorwf TMP,w
    andlw 0efh
    movwf portb
    bcf RS
    bsf E
    nop
    nop
    bcf E
    swapf tmp,w
    iorlw 0c0h
    andlw 0efh
    movwf portb
    bcf   RS
    bsf E
    nop
    nop
    bcf   E
    call p2
    return

wys_strona      movlw .0
          movwf adres
et_wys    call strona
          addlw 0
          btfsc status, z
          return
          call piszd
          incf adres, f
          movf adres, w
          goto et_wys
```

```
strona     addwf pcl
           dt "borowik.pl",0

wys_adres movlw .0
           movwf adres
et_wys1    call adres1
           addlw 0
           btfsc status, z
           return
           call piszd
           incf adres,f
           movf adres, w
           goto et_wys1
adres1     addwf pcl
           dt "borowik",0
    end
;------------------------
```

Program description

After clearing TRISB register and setting the internal generator frequency 4 MHz, the initLCD procedure is called. b

```
initLCD
    call p50
    movlw 0x13          ; E (RB4) = 1, data to be sent
                        ; (RB3 : RB0) = 3
                        ; w = 0001 0011
    movwf PORTB         ; w -> PORTB
    nop
```

```
nop
bcf E          ; the Enable strobe asserted low
call p5
bsf E          ; the Enable strobe asserted high
nop
nop
bcf   E
call p100
bsf E
nop
nop
bcf   E
call p100
movlw h'28' ; 4 bits mode, 2 lines, 5x7
call piszin
movlw .8          ; display off
call piszin
movlw .1          ; display clear
call piszin
movlw .6          ; entry mode
call piszin
movlw 0ch         ; display on
call piszin
return
```

Before we can use *initLCD*, the delay procedures *p50* (50 ms delay), *p5* (5 ms delay) and *p100* (100 us delay) must be defined. Also *initLCD* uses procedure for sending command to LCD: *piszin.* This procedure will be described later.
At the beginning *initLCD* procedure manually sends 3 times to LCD the value of 3 contained in the lower nibble of port B. Transfer is through raising and lowering the Enable strobe line.:

```
initLCD
    call p50          ; 50 ms delay
```

```
movlw 0x13          ; E (RB4) = 1, data to be sent
                    ; (RB3 : RB0) = 3
                    ; w = 0001 0011
movwf PORTB         ; w -> PORTB
nop
nop
bcf E        ; the Enable strobe asserted low
call p5
bsf E        ; the Enable strobe asserted high
nop
nop
bcf E        ; the Enable strobe asserted low
call p100
bsf E
nop
nop
bcf   E
call p100    ; 100 us delay
```

For transfering informations to LCD, the line E (Enable) is used. It is carried out with the sequence:

```
bsf E        ; the Enable strobe asserted high
nop
nop
bcf E        ; the Enable strobe asserted low
```

Raising and lowering voltage on the line is called *strobe*. It activates information transfer. Both the data and control lines are on the same port: port B, therefore care has to be taken, that line E is on the low state, otherwise any unwanted information can be sent to LCD. Usualy most of operation is performed on the working register W and, when the byte is ready to sent, it is splited into 2 nibbles, they are one after another moved to low 4 lines of port B and then the strobe signal is generated.

Next initialization sequence: sending commands: 0x28, 0x08, 0x01, 0x06 and 0x0c is implemented with the transfer procedure *piszin*, as shown below:

```
movlw h'28' ; 4 bits mode, 2 lines, 5x7
call piszin
```

```
movlw .8            ; display off
call piszin
movlw .1            ; display clear
call piszin
movlw .6            ; entry mode
call piszin
movlw 0ch           ; display on
call piszin
```

In binary it looks as below:

```
0 0 1 0  }  ( 28)  interfejs 4-bitowy,  2 linie
1 0 0 0

0 0 0 0  }  ( 08)  display off
1 0 0 0

0 0 0 0  }  ( 01)  display clear
0 0 0 1

0 0 0 0  }  ( 06)  entry mode set
0 1 1 0

0 0 0 0  }  ( 0C)  display on
1 1 0 0
```

Executing each command requires at least 1.6 ms. We provide after each transfer the 2 ms delay. The respective call (*call p2*) is included in the *piszin* procedure. The description of the *piszin* procedure, for transferring commands to LCD:

```
piszin              ; RS = 0
    movwf TMP
    bcf E           ; the Enable asserted low
```

```
swapf TMP,f
movlw 0c0h
iorwf TMP,w
andlw 0efh
movwf portb
bcf RS
bsf E
nop
nop
bcf E
swapf tmp,w
iorlw 0c0h
andlw 0efh
movwf portb
bcf   RS
bsf E
nop
nop
bcf   E
call p2
return
```

Before calling *piszin*, the byte to be sent is loaded to the working register W.
The LCD controller contains two separately addressable 8-bit registers: one for ASCII data and one for commands. LCD has an address line, RS, for the register selection. Its settings have the following meaning:

Register Select Control RS:

1 = LCD in data mode

0 = LCD in command mode.

First we assert the Enable strobe low,

```
bcf E                ; the Enable asserted low
```

The byte to be sent has two nibbles. We have to send higher nibble first. Send operation is made on the lower 4 lines of PORTB. Therefore the byte has to be swaped and its higher part is on the lower position. Next we have to clear bits 4th and 5th, so that lines RB4 (E signal) and RB5 (RS line) of port B will be cleared. We do this with *or* and *and* operations (commands: iorwf and andlw).

```
swapf TMP,f

movlw 0c0h

iorwf TMP,w

andlw 0efh

movwf portb
```

Example:
If we are to send command byte 0x28, we first load the value to variable register TMP. After swapf operation, TMP register will contain 0x82, binary b'1000 0010'. We have to make sure, that line 4 (Enable) is zero. Therefore we perform inclusive or operation of this value with the mask b'1100 0000', so that bits 6 and 7 are set, and lower bits are preserved without changes. Then we make and operation with the mask: b'1110 1111' and bit 4th is cleared. This looks as follows:

TMP	=	1000 0010	(0x82)
OR		1100 0000	
RESULT		1100 0010	; result goes to W
W	=	1100 0010	
AND		1110 1111	
RESULT		1100 0010	; result stays in W

Then the W value is moved to PORTB.
Now we manualy clear RS line:

```
bcf RS
```

and raise strobe line for three machine cycles:

```
bsf E

nop

nop

bcf E
```

Now it is time to send next nibble. TMP register is swapped again. After iorlw and andlw operations with tha same masks result is moved to PORTB:

```
swapf tmp,w
iorlw 0c0h
andlw 0efh
movwf portb
bcf   RS
bsf E
nop
nop
bcf   E
call p2
return
```

	TMP	=	0010 1000	(0x28)
OR			1100 0000	
RESULT			1110 1000	; result goes to W

	W	=	1110 1000	
AND			1110 1111	
RESULT			1110 1000	; result stays in W

At the very end, after transferring whole byte, the delay 2 ms is made.

```
call p2
```

After initialising LCD we have to address upper or lower line of LCD. LCD controller recognizes address command, if the MSB is set. Therefore the addres of the beginning of the upper LCD line would be not 0x00, but 0x80, binary 1 0 0 0 0 0 0 0. For the lower line addres starts with 0x40. After seting the MSB it would be in hex: 0xC0, binary 1100 0000.
Because those addresses are commands, RS line has to be held low.

```
        movlw 80h   ; address of the first charac-
ter, first line
        call piszin
```

We like to display in the upper line the string : borowik.pl. We generate this string and store in the memory with the procedure wys_strona.

```
wys_strona      movlw .0
                movwf adres
et_wys          call strona
                addlw 0
                btfsc status, z
                return
                call piszd
                incf adres, f
                movf adres, w
                goto et_wys

strona          addwf pcl
                    dt "borowik.pl",0
```

In the above procedures we form a lookup table. First we load 0 to the working register *W* and to the variable *address*. Then we call the one line procedure *strona*. Contents of *W* is table offset value. It is added to the Program Counter. Upon return W contains one by one all characters from the string. String is null terminated and after each return (retlw) contents of W is checked, if it is zero. If zero, procedure *wys_strona* ends. If not zero, character is sent to LCD in the procedure *piszd*, variable address determining table offset is incremented, is moved to W and again is called procedure *strona*.

dt directive (Define Table) generates a series of ***RETLW*** instructions, one instruction for each character in the string. Each character in a string is stored in its own ***RETLW*** instruction. This directive is used when generating a table of data for the PIC12/16 device family. If we were using a PIC18 device, it is recommended that we use the table read/write (***TBLRD***/***TBLWT***) features instead.

Let us again consider one line procedure *strona*

```
Strona      addwf pcl
            dt "borowik.pl",0
```

Reviewing program memory we can notice, that those lines are expanded as follows:

```
strona addwf   PCL  ,f
       retlw  0 x 62    ;  ASCII character 'b'
       retlw  0 x 6F    ;  ASCII character 'o'
```

```
retlw  0 x 72   ; ASCII character 'r'
retlw  0 x 6F   ; ASCII character 'o'
retlw  0 x 77   ; ASCII character 'w'
retlw  0 x 69   ; ASCII character 'i'
retlw  0 x 6B   ; ASCII character 'k'
retlw  0 x 2E   ; ASCII character '.'
retlw  0 x 70   ; ASCII character 'p'
retlw  0 x 6C   ; ASCII character 'l'
```

All those characters give the string "borowik.pl" that we see displayed on the LCD. Similarily in the second line there is displayed "borowik" (see procedure *wys_adres*).

Fig. .30 Strings displayed on LCD.

Procedure *piszin*, described earlier sends commands to the LCD controller. Procedure *piszd* for sending data (ASCII characters) to LCD is similar. The only difference is, that the RS (Register Select) line must be set high.

Test 13. Timer

Real Time Clock emulation on the LCD

Source code:

```
;LCD5 text demo - 4 bit mode
; RTC emulation

    LIST  p=16F628
    include "P16F628.inc"
    __config 0x3F10        ; configuration information

    cblock      0x20        ;constant block
        TMP
        a1     ; counter for delay procedure
        a2     ;     "           "          "
        a3     ;     "           "          "
        a4     ;     "           "          "
        g1     ; first digit of hours
        g2     ; second digit of hours
        m1     ; first digit of minutes
        m2     ; second digit of minutes
    endc
#define E portb,4
#define RS      portb,5
    org 0
    bsf   status, rp0
    movlw 0x00
    movwf trisb
    bsf   pcon, oscf  ; int. gen. 4 MHz
    bcf   status, rp0
    call initLCD
    movlw 0x30        ; ASCII code of '0' character
```

B. Borowik, *Interfacing PIC Microcontrollers to Peripherial Devices*, Intelligent Systems, Control and Automation: Science and Engineering 49, DOI 10.1007/978-94-007-1119-8_13,

```
    movwf g1     ; loading '0' character to g1
    movwf g2     ; loading '0' character to g2
    movwf m1     ; loading '0' character to m1
    movwf m2     ; loading '0' character to m2

    ; pok - procedure showing the time
    ; in the format: HH:MM (hours : minutes)

pok call go1     ; shows first digit of hours (g1)
    call go2     ; shows second digit of hours (g2)
    call mi1     ; shows first digit of minutes  (m1)
    call mi2     ; shows second digit of minutes (m2)
    incf m2,f    ; increments minutes (m2)
    movf m2,w          ; m2 -> W
    xorlw 0x3A         ; reached m2 10 minutes yet?
    btfsc status, Z    ; checking zero flag
    call n1            ; n1, if m2 reached ten minutes
    call pmin          ; wait 1 minute
    goto pok           ; shows new time

n1  movlw 0x30         ; W <- '0'
    movwf m2           ; m2 <- W   nbr of minutes
                       ; again 0
    incf m1, f         ; increment m1 (tens of
                       ; minutes)
    movf m1,w          ; m1 -> W
    xorlw 0x36         ; reached m1 6 (60 minutes)
                       ; yet?
    btfsc status, Z    ; Z flag
    call n2            ; n2, if m1 reached 6
    return
```

```
n2  movlw 0x30          ; W <- '0'
    movwf m1            ; m1 <- W   nbr of tens of
                        ; minutes again 0
    incf g2, f          ; increment g2 (hours)
    movf g2,w           ; g2 -> W
    xorlw 0x3A          ; reached g2 10 hours yet?
    btfsc status, Z     ; flaga Z
    call n3             ; n3, if g2 reached ten hours
    return

n3  movlw 0x30          ; W <- '0'
    movwf g2            ; g2 <- W   nbr of hours
                        ; again 0
    incf g1, f          ; increment g1
    movf g1,w           ; g1 -> W
    xorlw 0x32          ; reached g1 2 (20 hours)  yet?
    btfsc status, Z     ; flaga Z
    movlw 0x30          ; '0' ASCII
    movwf g1            ; '0' do g1
    return

initLCD
    call p50
    movlw 0x13
    movwf PORTB
    nop
    nop
    bcf E
    call p5
    bsf E
    nop
    nop
```

```
        bcf   E
        call p100
        bsf E
        nop
        nop
        bcf   E
        call p100
        movlw h'28'
        call piszin
        movlw .8
        call piszin
        movlw .1
        call piszin
        movlw .6
        call piszin
        movlw 0ch
        call piszin
        return

;pause 100 us
P100:  ;wait  t = 5 + 25*4 cycles
    movlw 0x01   ; 1 cycle
    movwf a1           ; 1 cycle
Out3:
    movlw 0x19   ; 1 cycle
    movwf 0x0E   ; 1 cycle
In3:
    decf 0x0E,1  ; 1 cycle
    btfss STATUS,Z           ; 1 cycle (Z = 0),
                             ; 2 cycles (Z = 1)
    goto In3             ; 2 cycles
    decf a1,1          ; 1 cycle
```

```
   btfss STATUS,Z              ; 1 cycle (Z = 0),
                               ; 2 cycles (Z = 1)
   goto Out3             ; 2 cycles
   return                ; 2 cycles
;-----------------------------


                          ;Pause 2 ms
P2:  ;wait   t = 4 + 10 * (6 + 50 * 4)cycles
   movlw  0x0A           ;1 cycle
   movwf  a1             ;1 cycle
Out2:
   movlw  0x32           ;1 cycle
   movwf  a2             ;1 cycle
In2:
   decf         a2,1           ;1 cycle
   btfss        STATUS,Z       ; 1 cycle (Z = 0),
                               ; 2 cycles (Z = 1)
   goto         In2            ; 2 cycles
   decf         a1,1           ; 1 cycle
   btfss        STATUS,Z       ; 1 cycle (Z = 0),
                               ; 2 cycles (Z = 1)
   goto         Out2           ; 2 cycles
   return                      ; 2 cycles
;-----------------------------


           ;Pause 5 ms
P5:        ;wait   t = 4 + 21 * (6 + 58 * 4)cycles
           ; = 4 + 21*238 = 4 + 4 998 us
   movlw  0x15           ; 1 cycle
   movwf  a1             ; 1 cycle
Out1:
   movlw  0x3A           ; 1 cycle
```

```
   movwf  a4             ; 1 cycle
In1:
   decf          a4,1          ; 1 cycle
   btfss         STATUS,Z      ; 1 cycle (Z = 0),
                               ; 2 cycles (Z = 1)
   goto          In1           ; 2 cycles
   decf          a1,1          ; 1 cycle
   btfss         STATUS,Z      ; 1 cycle (Z = 0),
                               ; 2 cycles (Z = 1)
   goto          Out1          ; 2 cycles
   return                      ;2 cycles
;-----------------------------


; Pause 50
P50:       ; wait   T = 10 * 5000 cycles
   movlw  0x0A           ; 1 cycle
   movwf  a2             ; 1 cycle
Jedziemy2:
   call          P5     ; 2 cycles
   decf          a2,1          ; 1 cycle
   btfss         STATUS,Z      ; 1 cycle (Z = 0),
                               ; 2 cycles (Z = 1)
   goto          Jedziemy2     ; 2 cycles
   return                      ; 2 cycles
    ;-----------------------------


piszd
    movwf TMP
    bcf E
    swapf TMP,f
    movlw 0c0h
    iorwf TMP,w
```

```
    andlw 0efh
    movwf portb
    bsf RS
    bsf E
    nop
    nop
    bcf E
    swapf tmp,w
    iorlw 0c0h
    andlw 0efh
    movwf portb
    bsf RS
    bsf E
    nop
    nop
    bcf   E
    call p100
    return

piszin          ; RS = 0
    movwf TMP
    bcf E
    swapf TMP,f
    movlw 0c0h
    iorwf TMP,w
    andlw 0efh
    movwf portb
    bcf RS
    bsf E
    nop
    nop
    bcf E
```

```
    swapf tmp,w
    iorlw 0c0h
    andlw 0efh
    movwf portb
    bcf   RS
    bsf E
    nop
    nop
    bcf   E
    call p2
    return

; Pause 1 min
; 60 times runs psec procedure

pmin:
   movlw 0x3c           ; 0x3c = 60
   movwf a4             ; a4 <- 60
Jedziemy7:
   call psec            ;2 cycles
   decf a4,1            ;1 cycle
   btfss STATUS,Z         ; 1 cycle (Z = 0),
                           ; 2 cycles (Z = 1)
   goto Jedziemy7        ;2 cycles
   return               ;2 cycles
;----------------------------

; psec procedure,
; waits 0.5 s, shows colon ':' on third position
; of LCD (address 0x82),
; again waits 0.5 s and replaces ':' with space ' '
psec      movlw 82h   ; third positions of
```

```
                        ; LCD address
    call piszin         ; send address to LCD
    movlw 0x3a          ; ':' ASCII

    call piszd          ; send ascii character to LCD

    ; 10 times calling p50 procedure
    ; 10 x 50 ms = 0.5 s
    movlw 0x0a          ; 10 -> W
    movwf a3            ; W -> a3
j4  call p50
    decf a3,1
    btfss status, Z
    goto j4

    ; again send the third positions of LCD address
    movlw 82h           ; third positions of
                        ; LCD address
    call piszin         ; send address to LCD

    movlw 0x20  ; ' ' (ascii code of space)
    call piszd

    ; calling the pause 50 ms 10 times
    movlw 0x0a  ; 10 -> W
    movwf a3    ; W -> a3
j5  call p50
    decf a3,1
    btfss status, Z
    goto j5
    return
```

```
; procedures go1, go2, mi1 i mi2 write time to LCD
; as HH:MM
; between hours and minutes procedure psec
; writes flickering colon
: (position 82h)
go1 movlw 0x80  ; LCD, first line, first character
    call piszin
    movf g1,w         ; g1 -> W
    call piszd
    return

go2 movlw 0x81  ; first line, second character
    call piszin
    movf g2,w         ; g2 -> W
    call piszd
    return

mi1 movlw 0x83  ; first line, 4th character
    call piszin
    movf m1,w         ; m1 -> W
    call piszd
    return

mi2 movlw 0x84  ; first line, 5th character
    call piszin
    movf m2,w         ; m2 -> W
    call piszd
    return

    end
;-----------------------------------------------
```

Program description

When microcotroller powers on, it starts counting the time and shows the time on the LCD display in the format HH:MM. Colon between hours and minutes is flickering with the frequency of 1 Hz. After 60 s number of minutes increases with 1. When the number of minutes reaches 10, it is cleared and tne number of tens of minutes is increased. When it reaches 6 (when it is 60 minutes), the number of hours increases and number of minutes is cleared and so on. Detailed explanation is in comment lines.

The application measures time correctly.

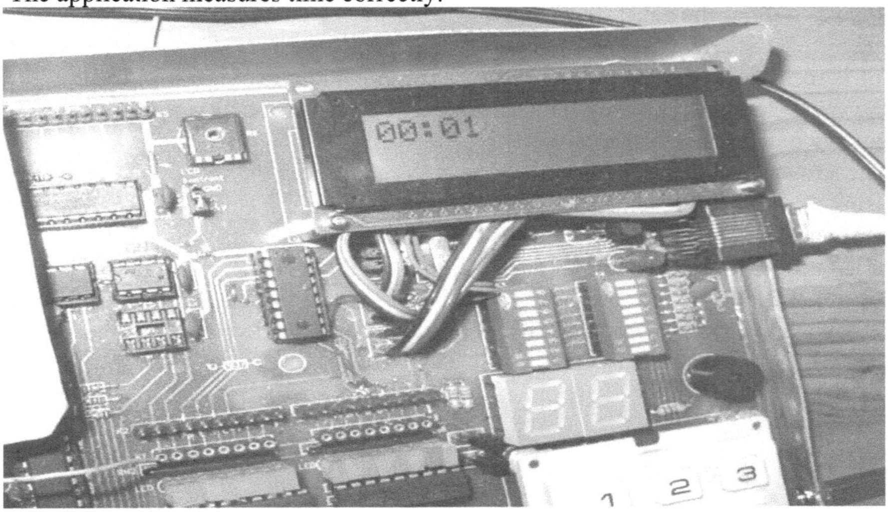

Fig. 31 Emulating the Real Time Clock. The colon blinks with the 1 Hz frequency.

Test 14. Dual RS232 software interface for PC and PIC microcontroller

In many PICMicros a built in serial interface, known as USART for "Universal \Synchronous/Asynchronous Receiver/Transmitter" is available. This hardware allows data to be received and sent without any software involvement. To send data, the USART simply shifts the data out (low byte first) to the TX I/O pin.
To receive data, the USART works by polling (sampling) the serial data input line at sixteen times the data rate. When a "low" is detected, the hardware waits 8 polling cycles to check to see if the line is still low half a bit period later. If it is, then a bit is read every sixteen polling clocks until the whole byte has been read in. At the end of the byte, the Stop bit is checked to be high. If the byte is received without any problems, a byte received flag is set.
The software presented in our test provides similar serial interface functions while using not USART module, but general purpose I/O pins. This type of interfacing is known as "Bit Banging". On the testing board serial input is connected to microcontroller pin 1 (PIC16F628 RA2). Serial output is connected to microcontroller pin (RA1). The Maxim "Max232" chip was used because it usesa single +5 Volt power supply. The schematic diagram is presented below.

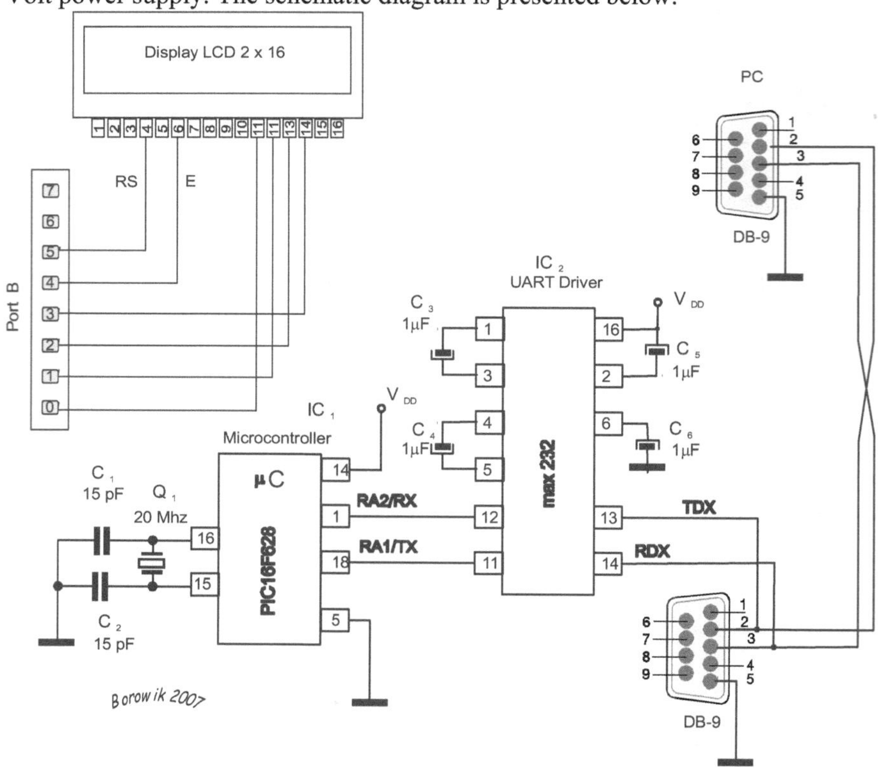

Fig. 32 Schemaic diagram for TX/RX operations

B. Borowik, *Interfacing PIC Microcontrollers to Peripherial Devices*, Intelligent Systems, Control and Automation: Science and Engineering 49, DOI 10.1007/978-94-007-1119-8_14,

Program source code:

```
; trans2.asm - demo application for the data
; transmission between PIC microcontroller and PC.
; ASCII values entered on the terminal window
; are transmitted by the RS232 link to the controller
; and displayed on the 2x16 LCD.
;The microcontroller sends feedback of received
; characters back to the issueing terminal window.
; Because we had troubles with using
; the built-in serial support USART,
; we designed our own code to implement
; serial communication.
; Such approach is called "bit banging".

    list p=16f628, r=hex     ; declare processor,
                             ; specifying the radix
    #include p16f628.inc     ; include register label
                             ; definitions
    __config h'3f10'  ; configuration information
                      ; for selected processor
                      ; 3f10 HVP (High Voltage
                      ; Programming),
                      ; 3f90 LVP (Low Voltage
                      ; Programming)

#define E portb,4     ; LCD on portB
#define RS      portb,5     ;     "     "

tmp equu  h'29'       ; auxiliary variable for
                      ; transmitting/receiving data
adres     equ   h'2E' ; variable for addressing
```

```
                              ; program memory
a1   equ    h'2F'             ; counter for delay procedures
a2   equ    h'30'             ;     "      "      "      "

a4   equ    h'32'             ;     "      "      "      "
b1   equ    h'33'             ; auxiliary variable for serial
                              ; communication
b2   equ    h'34'             ;     "             "            "

t6   equ    h'36'
tmp1        equ   h'38'
tmp3        equ   h'39'

WAIT:MACRO        TIME
            ; Delay for TIME µs.
            ; Variable TIME must be in multiples
            ; of 5 µs.
    MOVLW         (TIME/5) - 1        ; 1µs to process
    MOVWF         TMP1                ; 1µs to process
    CALL   WAIT5U                     ; 2µs to process
    ENDM

org 0
    movlw h'07'
    movwf cmcon
    bsf           status, rp0
    bcf           pcon, 1             ;Power On Reset
    movlw 0x00
    movwf trisb
    BCF           TRISA, 1
    movlw .4
    movwf trisa
```

```
    bsf   pcon, oscf  ; int.gen.4 MHz
    bcf   status, rp0
    movlw 0x20
    movwf fsr

    call  initLCD
    movlw 80h         ; LCD, first line
    call  piszin
    movlw .2
    movwf b2     ; counter
                 ; 2 times send string „boro"

att call  wys_micro
    ; procedure wys_micro reads characters from
    ; the string "boro" one after another,
    ; displays them on lcd
    ; and send them for transmission

    bsf   porta,1     ; RA1 on high (transmission in
                      ; idle state)
    call  p50         ; delay 50 ms
    call  p50         ;     "
    decfsz      b2,f
    goto  att         ; repeat it 2 times

; the text „boro" was transmitted to PC 2 times
; Then we enter ASCII characters on the terminal
; window on PC
; Microcontroller PIC receives them on the line RA2

at2 call  p50
    call  rs_rxd      ; data reception
```

```
    movwf t6            ; save received byte to
                        ; t6 variable
    call  piszd         ; display character on the lcd
    call  rs_tdx        ; echoing
; transmission of received characters
; back to the issueing terminal window

    call p50            ; delay for demo purposes
    call p50
    call p50
    call p50
    goto  at2           ; loop forever

initLCD                 ; LCD initialisation
    call  p50
    movlw 0x13    ; E (RB4) = 1, data to be sent
                  ; (RB3 : RB0) = 3
                  ; w = 0001 0011
    movwf PORTB         ; w -> PORTB
    nop
    nop
    bcf E         ; the Enable strobe asserted low
    call p5
    bsf E         ; the Enable strobe asserted high
    nop
    nop
    bcf   E
    call  p100
    bsf   E
    nop
    nop
    bcf   E
```

```
    call  p100
    movlw       h'28'        ; 4-bit mode, 2 lines
    call  piszin
    movlw       .8           ; display off
    call  piszin
    movlw       .1           ; display clear
    call  piszin
    movlw       .6           ; entry mode
    call  piszin
    movlw 0ch                ; display on
    call  piszin
    return

;pause 100 us
P100:                   ; wait t = 5 + 25*4 cycles
   movlw  0x01          ; 1 cycle
   movwf  a1            ; 1 cycle
Out3:
   movlw  0x19          ; 1 cycle
   movwf  a2            ; 1 cycle
In3:
   decf   a2,1                  ; 1 cycle
   btfss  STATUS,Z              ; 1 cycle (Z = 0),
                                ; 2 cycles (Z = 1)
   goto         In3             ; 2 cycles
   decf         a1,1            ; 1 cycle
   btfss        STATUS,Z        ; 1 cycle (Z = 0),
                                ; 2 cycles (Z = 1)
   goto         Out3            ; 2 cycles
   return                       ; 2 cycles
;---------------------------
                        ; pause 2 ms
```

```
P2:  ; wait t = 4 + 10 * (6 + 50 * 4)cycles
   movlw  0x0A          ;1 cycle
   movwf  a1            ;1 cycle
Out2:
   movlw  0x32          ;1 cycle
   movwf  a2            ;1 cycle
In2:
   decf   a2,1          ;1 cycle
   btfss        STATUS,Z     ; 1 cycle (Z = 0),
                             ; 2 cycles (Z = 1)
   goto         In2          ; 2 cycles
   decf         a1,1         ; 1 cycle
   btfss        STATUS,Z     ; 1 cycle (Z = 0),
                             ; 2 cycles (Z = 1)
   goto         Out2         ; 2 cycles
   return                    ; 2 cycles
;------------------------------


          ; pause 5 ms
P5:       ; wait t = 4 + 21 * (6 + 58 * 4)cycles
          ; = 4 + 21*238 = 4 + 4 998 us
   movlw  0x15          ; 1 cycle
   movwf  a1            ; 1 cycle
Out1:
   movlw  0x3A          ; 1 cycle
   movwf  a4            ; 1 cycle
In1:
   decf         a4,1         ; 1 cycle
   btfss        STATUS,Z     ; 1 cycle (Z = 0),
                             ; 2 cycles (Z = 1)
   goto         In1          ; 2 cycles
   decf         a1,1         ; 1 cycle
```

```
   btfss          STATUS,Z      ; 1 cycle (Z = 0),
                                ; 2 cycles (Z = 1)
   goto           Out1          ; 2 cycles
   return                       ; 2 cycles
;-----------------------------


;  pause 50 ms
P50:        ; wait T = 10 * 5000 cyclei
   movlw  0x0A          ; 1 cycle
   movwf  a2            ; 1 cycle
Jedziemy2:
   call           P5    ; 2 cycles
   decf           a2,1          ; 1 cycle
   btfss          STATUS,Z      ; 1 cycle (Z = 0),
                                ; 2 cycles (Z = 1)
   goto           Jedziemy2     ; 2 cycles
   return                       ; 2 cycles
    ;-----------------------------


piszd                   ; sending character to the lcd
    movwf TMP3
;   movwf t6
    bcf           E
    swapf         TMP3,f
    movlw         0c0h
    iorwf         TMP3,w
    andlw         0efh
    movwf         portb         ; high nibble transfer
    bsf           RS
    bsf           E
    nop
    nop
```

```
        bcf         E
        swapf       tmp3,w
        iorlw       0c0h
        andlw       0efh
        movwf       portb        ; low nibble transfer
        bsf         RS
        bsf         E
        nop
        nop
        bcf         E
        call  p100
        return

    piszin    ; RS=0, sending instruction to the lcd
        movwf       TMP3
        bcf         E
        swapf       TMP3,f
        movlw       0c0h
        iorwf       TMP3,w
        andlw       0efh
        movwf       portb
        bcf         RS
        bsf         E
        nop
        nop
        bcf         E
        swapf       tmp3,w
        iorlw       0c0h
        andlw       0efh
        movwf       portb
        bcf         RS
        bsf         E
```

```
    nop
    nop
    bcf         E
    call  p2
    return

WAIT5U:
                ; this takes 5 µs to complete
    NOP                 ; 1 µs to process
    NOP                 ; 1 µs to process
    DECFSZ      TMP1,F        ; 1 µs if not zero,
                              ; 2 µs if zero
    GOTO  WAIT5U        ; 2 µs to process
    RETLW       0

rs_tdx                        ; transmission routine
    movf        t6,w          ; w <- t6
    movwf       tmp
    movlw       .9      ; 8 data bits + 1 stop bit
                        ; (8 + 1 = 9)
    movwf       b1
    bsf         status,C
    bcf         porta, 1
rs_tx1
    call        p5
    call        p2
    call        p2
    wait        .50           ; wait 50 µs

          ; total: 5 + 2 + 2 + 0,050 = 9,05 ms
    rrf         tmp, f        ; right shift
    btfsc       status,c      ; check the state
```

```
                                ; of C flag, if it is 0
     goto        s1             ; it is high
     goto        c1             ; it is low
s1   bsf         porta, 1       ; bring RA1 (transmit)
                                ; high
     goto        d1
c1   bcf         porta,1        ;bring RA1 (transmit) low
d1   decfsz      b1, f
     goto        rs_tx1

                                ; long delay after
                                ; transmitting the byte
     call        p5
     call        p2
     call        p2
     wait        .50
     call        p5
     call        p2
     call        p2
     wait        .50
     return

rs_rxd                          ; reception routine
     movlw       .8             ; 8 bits to be received
     movwf       b1
     btfsc       porta,2        ; polling the RA2 line
                          ; until the leading edge
                          ; of the Start bit is detected
     goto        rs_rxd

          ; delay of 2 + 2 + 0,520 = 4,52 ms,
          ; approximately ½ of the bit time
```

```
    call        p2
    call        p2
    wait        .520
rs_rx1
    ; sampling the middle of the each of the 8 bits
    ; of data after delaying 9.05 ms (app. bit time)
    call        p5
    call        p2
    call        p2
    wait        .50             ; macro for delay 50 us

    bcf         status, C       ; clearing C flag
    btfsc       porta, 2        ; is RA2 input line low
                                ; or high?
    bsf         status, C       ; if high, C = 1
                                ; otherwise C remains 0
    rrf    tmp, f          ; rotate right variable tmp
                                ; through Carry Flag

    decfsz      b1, f           ; all bits received yet?
    goto        rs_rx1          ; receiving next bit

    movf        tmp, t6         ; store received byte
                                ; to t6
    movf        tmp, w          ; and to W
    return

wys_micro                  ; reading lookup table
                ; displaying the character on LCD
                ; and sending it to transmission
    movlw       .0
    movwf       adres
```

```
et_w
    call        napis        ; call lookup table

    addlw       .0     ; if the text terminated,
                       ; the Z flag will be set
    btfsc       status,z
    return
    movwf t6           ; store the read character
                       ; to t6 variable
    call        piszd        ; character sent to lcd
    call        rs_tdx       ; character sent
                             ; for transmission
    incf        adres, f     ; next character
    movf        adres, w
    goto        et_w

    napis addwf pcl, f       ; Add offset to table
                             ; base pointer
    dt     " boro ", 0       ; create table

;----------------------------------------------
  end
```

Program description
We use Bit-Banging method to emulate serial port. In this method we manage all synchronization and timing signals and we have to ensure proper setup and times for reading and writing data to the ports.
Bits are transfered in „8-N-1" format which means that eight data bits are sent with a single (low) Start Bit and high Stop Bit to make up a data packet. "8-N-1 is a simple and convenient protocol. The packet ends with a "1" being sent as a Stop Bit, to allow the "0" Start Bit to be easily recognised. In the idle state the line is on high. N letter means that no parity checking is performed.
Each bit is transmitted for a set period of time, the reciprocal of which is the "data rate". Common data rates are 300, 1200, 2400, 9600, 19200 bps. As equipment

got faster, the data rates became multiples of these speeds. For the rate 9600 bps the bit time is:

$$1/9\,600 = 104.1\ \mu s$$

In our experiment we use the slow transmission: 110 bps. For this rate the bit time is:

$$1/110 = 9.09\ ms$$

The signal for each bit, whether 1 or 0, has to stay on the transmit line over the time of app. 9.09 ms. We achieve this delay adding up: 2ms+ 2ms + 5ms + 50 μs = 9.05 ms. The least significant bit is sent first, immediately after the start bit that is always low (0).

The start bit is an initial zero bit and the stop bit is a trailing one bit. The binary code b'01000001' (code of the character 'A') is actually transmitted as follows: first the start bit 0, then 8 bits in reversed order and the stop bit high: : -> 0,0,1,0,1,0,0,1,0,1 with the start and stop bits added and the least significant bit sent first. The signal line idles at a '1' (high) normally so the start bit is a change from idle to a 0 and the stop it continues at a one.

It is good custom to end the tranmission with 2 stop bits (8N2), while perform reception with the format 8N1. It provides some synchronization between TX and RX, assuming approximations in the delay procedures. Also between bytes we entered longer delays for the demo purposes.

To receive data, the input line is continuously polled. When the leading edge of the start bit is detected, a half bit delay is made:

$$9{,}09/2 = 4{,}5\ ms.$$

and the input line is polled again 8 times with time intervals 9.09 ms after each poll. At the end of the packet the Stop Bit is checked and the routine returns to its caller with the received byte.

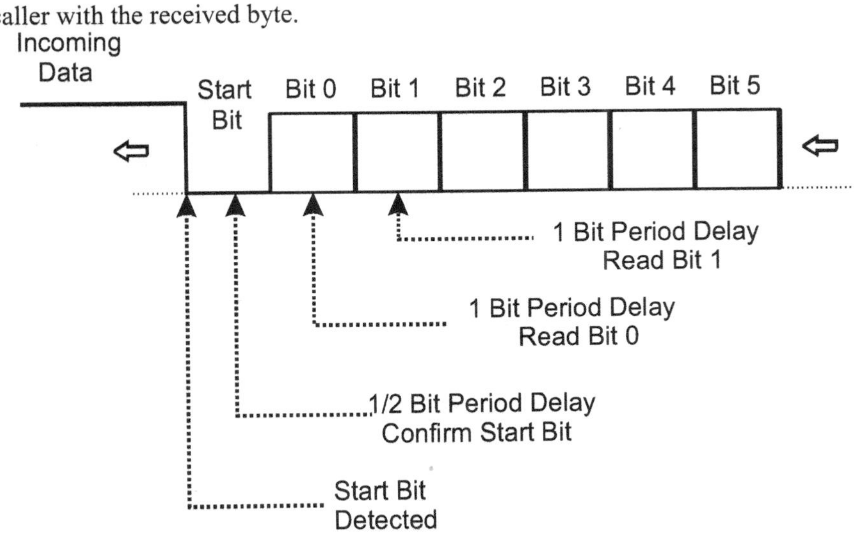

Fig. 33 Bit banging rs-232 data receive

For measuring delay time we used 5 procedures and one macro:

wait5u	delay	5 µsek
p100	„	100 µs
p2	„	2 ms
p5	„	5 ms
p50	„	50 ms
wait: time.	Delay for TIME µs (macro).	

Setting signal low or high on the TX line (RA1) uses the following technique: Byte, to be sent is stored in the auxiliary variable *tmp*.

```
movf        t6,w        ; w <- t6
movwf       tmp
```

First we like to send Start Bit, therefore RA1 is put on low and bit time 9.05 ms is delayed.

```
bcf         porta, 1
....
call        p5
call        p2
call        p2
wait        .50         ; wait 50 µs
            ; total: 5 + 2 + 2 + 0,050 = 9,05 ms
```

Now the variable *tmp* is rotated right through Carry. The most right bit (less significant) is loaded to Carry bit of the STATUS register.

```
rrf         tmp, f      ; rotate right tmp
                        ; through Carry
```

Then the state of C flag is checked. If it is high, the line RA1 is set high too. If C flag is low, RA1 is brought to low. Finaly the delay 9.05 ms is performed. This procedure is repeated 9 times (counter b1), so, that the whole byte and stop bit are outputed. Last bit sent is '1', because before transmission the C Flag was set.

```
movlw       .9      ; 8 data bits + 1 stop bit
                    ; (8 + 1 = 9)
movwf       b1
```

```
        bsf         status,C      ; setting C high before
                                  ; transmission

        ...
        rrf         tmp, f        ; rotate right tmp
                                  ; through Carry
        btfsc       status,c      ; check the state
                                  ; of C flag, if it is 0
        goto        s1            ; it is high
        goto        c1            ; it is low
s1      bsf         porta, 1      ; bring RA1 (transmit)
                                  ; high
        goto        d1
c1      bcf    porta,1      ; bring RA1 (transmit) low
d1      decfsz      b1, f         ; decrementing counter
        goto        rs_tx1
```

After sending byte, the long delay (two bit times) is performed.

```
        ; long delay after transmitting the byte
        call        p5
        call        p2
        call        p2
        wait        .50
        call        p5
        call        p2
        call        p2
        wait        .50
        return
```

Receiving data is carried out in the similar manner (procedure rs_rxd).
The state if RA2 line is checked. If it is high, Carry flag is set high. If RA2 is low, the Carry Flag is brought low too. Then the auxiliary variable *tmp* is rotated right through Carry, loading with received bits.

```
    . . . .
rs_rx1
; sampling the middle of the each of the 8 bits
; of data
; after delaying 9.05 ms (app. bit time)
    call        p5
    call        p2
    call        p2
    wait        .50         ; macro for delay 50 us

    bcf         status, C   ; clearing C flag
    btfsc       porta, 2    ; is RA2 input line low
                            ; or high?
    bsf         status, C   ; if high, C = 1
                            ; otherwise C remains 0
    rrf         tmp, f      ; rotate right
                            ; variable tmp
                            ; through Carry Flag

    decfsz      b1, f       ; all bits received yet?
    goto        rs_rx1      ; receiving next bit

    movf        tmp, t6     ; store received byte
                            ; to t6
```

Above presented application makes first reading the text 'boro\ from the lookup table. Text is 2 times written to lcd and transmitted to terminal window on PC. It can be Microsoft HyperTerminal application, but we would recommend application Tera Term Pro because of its simplicity.

After displaying on the terminal window the text: boro boro, the transmission from the PIC microcontroller is terminated and the microcontroller waits for data from PC. ASCII values entered on the terminal window are transmitted by the RS232 link to the controller and displayed on the LCD. The microcontroller sends feedback of received characters back to the issueing terminal window.

Test 15. Matrix Keypad + serial transmission

Keypads play important role in a small embedded system where human interaction or human input is needed. All we have to do is connect the rows and columns to a port of microcontroller and program the controller to read the input.

We make the rows as i/p and we drive the columns making them o/p, this whole procedure of reading the keyboard is called scanning.

A 3 x 4 matrix keypad is used for data entry. The 12 push switches are connected to seven lines of port A, as shown in figure 36.

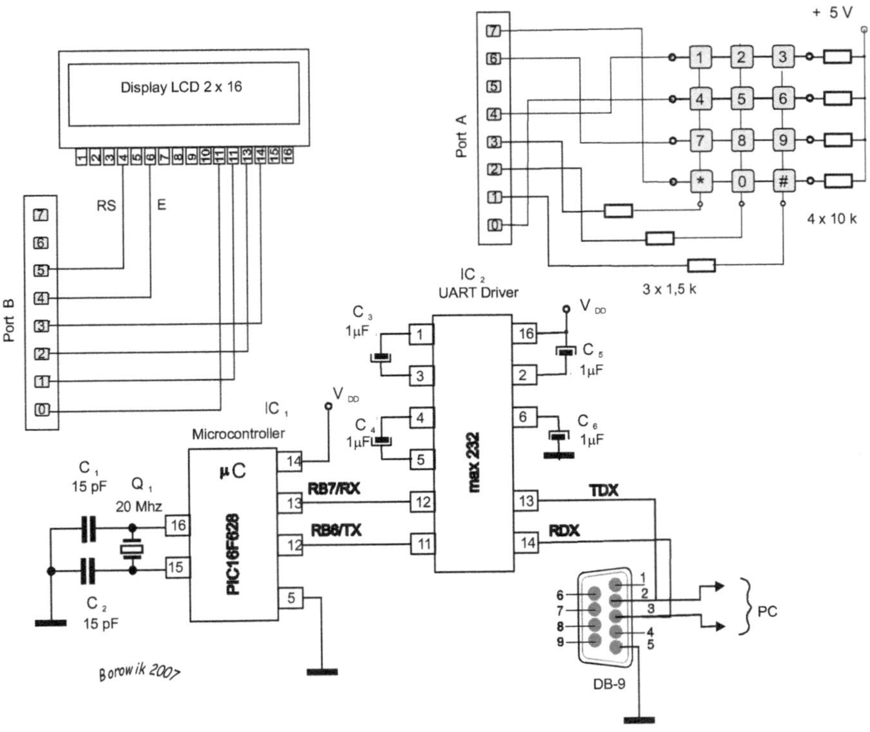

Fig. 34 Schematic diagram for the test with the matrix keypad

Port A and Port B lines are connected as follows:

RA0	keypad	Row 2	RB0	LCD	Data
RA1		Col 3	RB1		

B. Borowik, *Interfacing PIC Microcontrollers to Peripherial Devices*, Intelligent Systems, Control and Automation: Science and Engineering 49, DOI 10.1007/978-94-007-1119-8_15,

RA2		Col 2	RB2		
RA3		Col 1	RB3		
RA4		Row 1	RB4		Enable
RA5	Not used		RB5		Register Select
RA6	Keypad	Row 3	RB6	RS-232	TX
RA7		Row 4	RB7		RX Not used in the program

Or with reference to modules:

Matrix keypad:

Row 1	RA4
Row 2	RA0
Row 3	RA6
Row 3	RA7
Col 1	RA3
Col 2	RA2
Col 3	RA1

LCD:

Data	<RB0 : RB3>
Enable	RB4
RS	RB5

Serial transmission:

TX	RB6
RX	RB7

Program source code

```
;
    list p=16f628, r=hex      ; declare processor,
                              ; specifying the radix
    #include p16f628.inc      ; include register label
                              ; definitions
    __config h'3f10'  ; configuration information for
                      ; selected processor
                ; 3f10 HVP (High Voltage Programming)
                 ; 3f90 LVP (Low Voltage Programming)
#define E portb,4             ; LCD on portB
#define RS      portb,5       ;      "      "

tmp         equ   h'29'
cur_col     equ   h'2A'       ; current column
key_val     equ   h'2C'       ; ASCII code of the
                              ; pressed key


a1  equ   h'2F'
a2  equ   h'30'
wait_count      equ   h'31'
a4  equ   h'32'
b1  equ   h'33'
b2  equ   h'34'
tmp2        equ   h'35'
t6  equ   h'36'
tmp1        equ   h'38'
tmp3        equ   h'39'
w_tmp equ h'3a'
st_tmp equ      h'3b'

l_msek    equ   h'3d' ; delay counter
```

```
l_kol           equ   h'3e' ; column counter
NO_HIT    equ   0F1h        ; the mask for port A
                            ; if no hit: 1111 0001


WAIT:MACRO      TIME
          ; Delay for TIME µs.
    ; Variable TIME must be in multiples of 5 µs.
    MOVLW       (TIME/5) - 1      ; 1µs to process
    MOVWF       TMP1              ; 1µs to process
    CALL  WAIT5U                  ; 2µs to process
    ENDM
;;;;;;;;;;;;;;;;;;;;;;;;;;;;;;;;;;;;;;;;;;;;;;;;;
org 0
    goto _start

org 4
_int      ; interrupt service routine
          ; not used in this program

    ; store the contents of W and STATUS registers
    movwf w_tmp       ; w_tmp <- w
    swapf status, w   ; w <- status
    movwf st_tmp      ; st_tmp <- w

    bcf status, rp0   ; bank 0
    btfsc intcon, t0if
    goto  _t1
    goto _odtworz

_t1 bcf   intcon, T0IF      ; clear flag overflow
                      ; TMR0 to be able to react
                      ; to the next interrupt
```

```
    movlw .6                    ; load TMR0
    movwf TMR0

    movfw wait_count
    btfss status, Z
    decf  wait_count, f
    goto  _odtworz

    ; restoring registers
_odtworz  swapf st_tmp, w   ; w <- st_tmp
    movwf status        ; status <- w
    swapf w_tmp, f      ; w <- w-tmp
    swapf w_tmp, w
    retfie
**************************************************
; main
;*****************************************

_start
    movlw h'07'
    movwf cmcon
    bsf   status, rp0
    bcf         pcon, 1             ;Power On Reset
    movlw 0x00
    movwf trisb
    bsf   pcon, oscf                ; int.gen.4 MHz

    movlw .1            ; prescaler is assigned
                        ; to the Timer0 module
    movwf option_reg    ; prescaler rate:  1:4

    movlw       0d1h    ; portA direction b'1101 0001'
```

```
; RA0 Input, RA1:RA3 Output, RA4 Input, RA5 Output,
                        ; RA6:RA7 Input
    movwf trisa

    bcf   status, rp0
    movlw 000h          ; 0 to portA and portB
    movwf porta
    movwf portb

    clrf  intcon        ; clear intcon, clear rbif flag
                        ; disabling all interrupts

    movlw .6            ; starting value of TMR0
    movwf tmr0

    call initLCD
    movlw 80h   ; first line of the LCD
    call piszin

    clrf  cur_col

    clrf  key_val
    clrf  wait_count

    ;bsf  intcon, t0ie       ;enables the TMR0
                             ; interrupt
    ;bsf  intcon, gie        ;enables all interrupts

; end of the initialisation
;;;;;;;;;;;

scan                ; reading the keypad
```

```
    incf  cur_col, f         ; pickup next column
                             ; to scan
    movlw .4            ; if cur_col > max_col, then
    subwf  cur_col, w        ; we just did last column
                             ; so start over
    btfss status, z     ; if zero we need to start over
    goto  key_scan      ; if not, go, look for key hits
    clrf  cur_col       ; starting over
    goto  scan

key_scan                ; reading the key
    movfw cur_col       ; get bit pattern, which
    call  col_select    ; selects currently
                        ; desired column
    movwf porta         ; and enable that column

    movfw porta         ; reading rows (input lines)
    andlw 0xd1          ; 0xd1 = b1101 0001, porta
                        ; if no key hit
                        ; clearing column lines <1:3>
    addlw -NO_HIT       ; see, if key hit occurred
                        ; in current column
    btfsc status, z
    goto  scan          ; no, look at next column
    call key_get        ; yes, process ignore
                        ; key release keystroke

    andlw 0d1h          ; there is bounce on
                        ; a key „release"
    xorlw 0d1h

    btfsc status, z     ; we want to debounce
```

```
                            ; key releases
    goto  scan              ;(so, they do not look like
                            ; legit keystrokes) but then we
                            ; just ignore the release

    call  key_xlate         ; legit keystroke - turn into
                            ; binary value

    movfw key_val
    addlw .35               ; ASCII code of '#' character
    movwf t6
    call  piszd

    bsf   portb, 6          ;HIGH on RB6 (transmitting)
    call rs_tdx
    call p50
    call p50
    call p50
    call p50
    goto scan     ; scan forever

initLCD
    call  p50
    movlw 0x13              ; E (RB4) = 1, data to be sent
                            ; (RB3 : RB0) = 3
                            ; w = 0001 0011
    movwf PORTB             ; w -> PORTB
    nop
    nop
    bcf E         ; the Enable strobe asserted low
    call p5
    bsf E         ; the Enable strobe asserted high
```

```
    nop
    nop
    bcf   E
    call p100
    bsf E
    nop
    nop
    bcf   E
    call p100
    movlw h'28' ; 4 bits mode, 2 lines, 5x7
    call piszin
    movlw .8          ; display off
    call piszin
    movlw .1          ; display clear
    call piszin
    movlw .6          ; entry mode
    call piszin
    movlw 0ch         ; display on
    call piszin
    return

col_select
    addwf pcl, f      ; get bit pattern, which
                      ; selects given column
    dt    b'00001110'       ; no column selected
    dt    b'00000110'       ; col 1
    dt    b'00001010'       ; col 2
    dt    b'00001100'       ; col 3

key_get
    movfw porta
    movwf key_val
```

```
check
    call  p5
    call  p5
    call  p5
    call  p5
    movfw porta
                        ; debounce cycle
    subwf key_val, f    ; see, if matches last
    btfsc status, z     ; if Z, the values matched
    goto matched
    movwf key_val       ; no match, start debounce
                        ; cycle again

    goto  check

matched
    movwf key_val       ; save in case, we have to do
                        ; again
    return

key_xlate
    comf  key_val, w    ; complement so only bits on
                        ; correspond
    andlw 0d1h          ; to row select mask unused off

    call  bit2row       ;translate it into a row number
    movwf key_val

    movfw cur_col       ; get current column
    addlw -1      ; convert into an 0-relative offset
    addwf pcl, f
```

```
    goto  col1_xlate
    goto  col2_xlate
    goto  col3_xlate

col1_xlate:
    movfw key_val
    call col1_keys
    movwf key_val
    return

col2_xlate:
    movfw key_val
    call col2_keys
    movwf key_val
    return

col3_xlate:
    movfw key_val
    call col3_keys
    movwf key_val
    return

col1_keys:
    addwf pcl, f
    dt    .14          ; character ,1'
    dt    11h          ; character ,4'
    dt .20             ; character ,7'
    dt .7              ; character ,*'

col2_keys:
    addwf pcl, f
    dt    0fh          ; character ,2'
```

```
    dt     .18          ; character ,5'
    dt     .21          ; character ,8'
    dt     .13          ; character ,0'

col3_keys:
    addwf pcl, f
    dt     10h          ; character ,3'
    dt     .19          ; character ,6'
    dt     .22          ; character ,9'
    dt     .0           ; character ,#'

bit2row:
    movwf tmp
    sublw 0x10
    btfss status, Z
    goto r2
    retlw .0

r2  movfw tmp
    sublw .1
    btfss status, Z
    goto r3
    retlw .1

r3  movfw tmp
    sublw .64
    btfss status, Z
    goto r4
    retlw .2

r4  movfw tmp
    sublw .128
```

```
    btfss status, Z
    return
    retlw .3

; pause 100 us
P100:  ; wait t = 5 + 25*4 cycles
   movlw 0x01          ; 1 cycle
   movwf a1                  ; 1 cycle
Out3:
   movlw 0x19          ; 1 cycle
   movwf a2                  ; 1 cycle
In3:
   decf a2,1           ; 1 cycle
   btfss STATUS,Z            ; 1 cycle (Z = 0),
                             ; 2 cycles (Z = 1)
   goto In3                  ; 2 cycles
   decf a1,1           ; 1 cycle
   btfss STATUS,Z            ; 1 cycle (Z = 0),
                             ; 2 cycles (Z = 1)
   goto Out3             ; 2 cycles
   return                ; 2 cycles
;-----------------------------

; pause 2 ms
P2:  ; wait t = 4 + 10 * (6 + 50 * 4)cycles
   movlw 0x0A          ; 1 cycle
   movwf a1                  ; 1 cycle
Out2:
   movlw 0x32          ; 1 cycle
   movwf a2                  ; 1 cycle
In2:
   decf a2,1             ;1 cycle
```

```
    btfss STATUS,Z              ; 1 cycle (Z = 0),
                                ; 2 cycles (Z = 1)
    goto In2                    ; 2 cycles
    decf a1,1             ; 1 cycle
    btfss STATUS,Z              ; 1 cycle (Z = 0),
                                ; 2 cycles (Z = 1)
    goto Out2               ;2 cycles
    return                  ;2 cycles
;----------------------------


; pause 5 ms
P5:  ; wait t = 4 + 21 * (6 + 58 * 4)cycles
   ; = 4 + 21*238 = 4 + 4 998 us
   movlw 0x15          ; 1 cycle
   movwf a1                  ; 1 cycle
Out1:
   movlw 0x3A          ; 1 cycle
   movwf a4                  ;1 cycle
In1:
   decf a4,1           ; 1 cycle
   btfss STATUS,Z            ; 1 cycle (Z = 0),
                             ; 2 cycles (Z = 1)
   goto In1                  ; 2 cycles
   decf a1,1           ; 1 cycle
   btfss STATUS,Z            ; 1 cycle (Z = 0),
                             ; 2 cycles (Z = 1)
   goto Out1              ; 2 cycles
   return                 ; 2 cycles
;----------------------------


;  pause 50 ms
P50:      ; wait T = 10 * 5000 cycles
```

```
   movlw 0x0A          ; 1 cycle
   movwf a2                  ; 1 cycle
Jedziemy2:
   call P5                   ; 2 cycles
   decf a2,1           ; 1 cycle
   btfss STATUS,Z            ; 1 cycle (Z = 0),
                             ; 2 cycles (Z = 1)
   goto Jedziemy2            ; 2 cycles
   return              ; 2 cycles
    ;-----------------------------

piszd
     movwf TMP3
     bcf E
     swapf TMP3,f
     movlw 0c0h
     iorwf TMP3,w
     andlw 0efh
     movwf portb        ; high nibble transfer
     bsf RS
     bsf E
     nop
     nop
     bcf E
     swapf tmp3,w
     iorlw 0c0h
     andlw 0efh
     movwf portb        ; low nibble transfer

     bsf RS
     bsf E
     nop
```

```
    nop
    bcf   E
    call p100
    return

piszin          ; RS = 0
    movwf TMP3
    bcf E
    swapf TMP3,f

    movlw 0c0h
    iorwf TMP3,w
    andlw 0efh
    movwf portb
    bcf RS
    bsf E
    nop
    nop
    bcf E
    swapf tmp3,w
    iorlw 0c0h
    andlw 0efh
    movwf portb
    bcf   RS
    bsf E
    nop
    nop
    bcf   E
    call p2
    return

WAIT5U:               ; This takes 5 µs to complete
```

```
    NOP                  ; 1 µs to process
    NOP                  ; 1 µs to process
    DECFSZ      TMP1,F        ; 1µs if not zero or
                              ; 2 µs if zero

    GOTO  WAIT5U         ; 2 µs to process
    RETLW       0

rs_tdx
    movf t6,w
    movwf tmp
    movlw .9
    movwf b1
    bsf         status,C
    bcf   portb, 6
rs_tx1
    call p5
    call p2
    call p2
    wait .50
    rrf   tmp, f
    btfsc status,c
    goto  s1
    goto c1
s1  bsf   portb, 6
    goto d1
c1  bcf   portb,6
d1  decfsz      b1, f
    goto  rs_tx1
    call p5
    call p2
    call p2
```

```
    wait .50
    call p5
    call p2
    call p2
    wait .50
    return

;---------------------------------------------
  end
```

Program description
Interrupt Service Routine has not been used in the program.
The three columns of the keypad are connected to three of the output signals from the PIC. The four rows of the keypad are connected to four of the input signals to the PIC. If no key is pressed there is no electrical contact between the rows and the columns. The PIC detects which key is pressed by 'scanning' the keypad.
First the PIC makes the column 1 current column and sets the signal going to Column 1 low (a digital '0'). All the other Columns are made high (a digital '1'). Then it tests the input signals coming from each Row in turn. If any of the Row signals is low, this means that the corresponding key in Column 1 has been pressed. Current column number is stored to the variable *cur_col.*
Interrupt Service Routine has not been used in the program. During the initialization phase the variable *cur_col* is cleared:

```
clrf  cur_col
```

After modules initialization the keypad scanning is executed.

```
scan                   ; reading the keypad
    incf  cur_col, f   ; pickup next column to scan
    movlw .4           ; if cur_col > max_col, then
    subwf  cur_col, w        ;we just did last column
                             ; so start over
    btfss status, z    ;if zero we need to start over
    goto  key_scan     ;if not, go, look for key hits
    clrf  cur_col      ;starting over
    goto  scan
```

```
key_scan                ; reading the key
    movfw cur_col       ; get bit pattern, which
    call  col_select    ; selects currently desired
                        ; column
    movwf porta         ; and enable that column
```

We call *col_select* to receive the bit pattern for port A with the specified column selected.

```
col_select      addwf pcl, f        ; get bit pattern
                        ;, which selects given column
    dt    b'00001110'       ; no column selected
    dt    b'00000110'       ; col 1, bit 3 low
    dt    b'00001010'       ; col 2, bit 2 low
    dt    b'00001100'       ; col 3, bit 1 low
```

Then we send received bit pattern to port A and respective column output line is set low. Next we read port A to detect, if any row is made low on input row lines. Such situation means, that the corresponding key has been pressed.

```
    movwf porta         ; and enable that column
    ...
    movfw porta         ; reading rows (input lines)
```

We have to compare the setting of port A with the pattern NO_HIT (No key pressed) and to debounce key releases.

```
    movfw porta         ; reading rows (input lines)
    andlw 0xd1          ; 0xd1 = b1101 0001, porta
                        ; if no key hit
                        ; clearing column lines <1:3>
    addlw -NO_HIT       ; see, if key hit occurred
                        ; in current column
    btfsc status, z
```

```
    goto  scan           ; no, look at next column
    call key_get         ; yes, process ignore key
                         ; release keystroke

    andlw 0d1h           ; there is bounce on a key
                         ; „release"
    xorlw 0d1h

    btfsc status, z      ; we want to debounce key
                         ; releases
    goto  scan           ;(so, they do not look like
                         ; legit keystrokes) but then we
                         ; just ignore the release
    ...
key_get
    movfw porta
    movwf key_val

check
    call  p5
    call  p5
    call  p5
    call  p5
    movfw porta
                         ; debounce cycle
    subwf key_val, f     ; see, if matches last
    btfsc status, z      ; if Z, the values matched
    goto matched
    movwf key_val        ; no match, start debounce
                         ; cycle again

    goto  check
```

```
matched
    movwf key_val       ; save in case, we have to do
                        ; again
    return
```

Now if we know valid keys, they have to be translated into ASCII codes. This task carries out the routine *key_xlate*. The lookup tables are built with the respective values for each column and each row. The lowest ASCII code is for the character # (ASCII code 35)In lookup tables we assign 0 for the # character and starting with 0 we successively assign next numbers for next characters. Therefore after finishing this procedure to each value we have to add 35.

```
key_xlate
    comf  key_val, w    ; complement so only bits on
                        ; correspond
    andlw 0d1h          ; to row select mask unused off

    call  bit2row       ;translate it into a row number
    movwf key_val

    movfw cur_col       ; get current column
    addlw -1            ; convert into an 0-relative
                        ; offset
    addwf pcl, f
    goto  col1_xlate
    goto  col2_xlate
    goto  col3_xlate

col1_xlate:
    movfw key_val
    call col1_keys
    movwf key_val
```

```
    return

col2_xlate:
    movfw key_val
    call col2_keys
    movwf key_val
    return

col3_xlate:
    movfw key_val
    call col3_keys
    movwf key_val
    return

col1_keys:
    addwf pcl, f
    dt    .14          ; character ,1'
    dt    11h          ; character ,4'
    dt .20             ; character ,7'
    dt .7              ; character ,*'

col2_keys:
    addwf pcl, f
    dt    0fh          ; character ,2'
    dt    .18          ; character ,5'
    dt    .21          ; character ,8'
    dt    .13          ; character ,0'

col3_keys:
    addwf pcl, f
    dt    10h          ; character ,3'
    dt    .19          ; character ,6'
```

```
    dt    .22          ; character ,9'
    dt    .0           ; character ,#'

bit2row:
    movwf tmp
    sublw 0x10
    btfss status, Z
    goto r2
    retlw .0

r2  movfw tmp
    sublw .1
    btfss status, Z
    goto r3
    retlw .1

r3  movfw tmp
    sublw .64
    btfss status, Z
    goto r4
    retlw .2

r4  movfw tmp
    sublw .128
    btfss status, Z
    return
    retlw .3
```

Finaly the found ASCII codes are sent to LCD for displaying and in parallel for transmission to terminal window on PC.

The Stack Memory

During tracing the software execution very helpful is watching the memory contents.

In the PIC mocrocontrollers we can distinguish the program memory and the data memory. There is also the stack memory, which is not space of either program or data memory.

The PIC 18F1320 has 8 Kbytes (0x2000) of Flash program memory and can store up to 4096 (0x1000) single-word instructions. The Reset vector address is at 0000h. The program memory map is shown in the figure below.

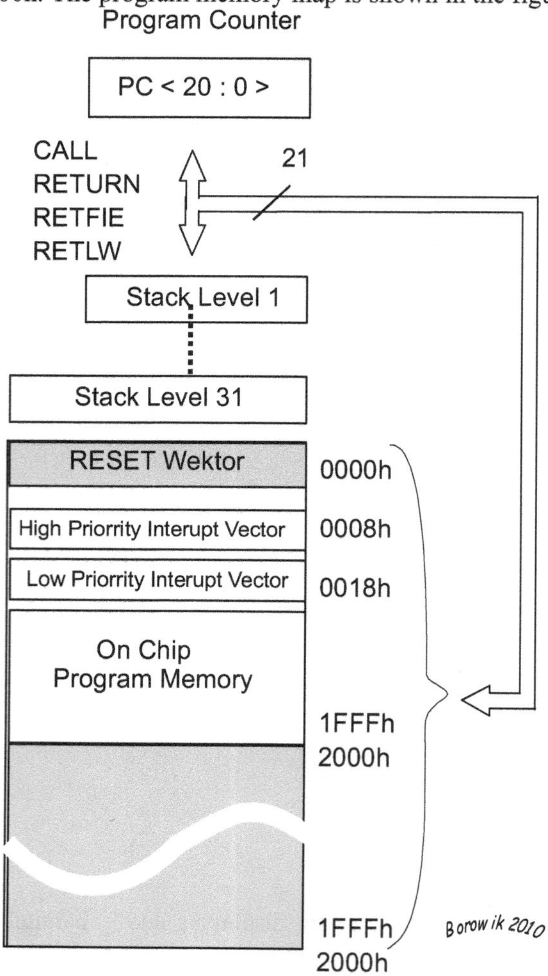

Fig.35. Program memory map and stack for PIC18F1320

B. Borowik, *Interfacing PIC Microcontrollers to Peripherial Devices*, Intelligent Systems, Control and Automation: Science and Engineering 49, DOI 10.1007/978-94-007-1119-8,

The stack allows up to 31 program calls and interrupts to occur. The PC register content is pushed onto the stack, when a call instruction is executed. The PC value is pulled off the stack on a RETURN, RETLW, or RETFIE instruction.

The stack operates as a 31 word by 21 bit RAM and a 5 bit stack pointer, initialized to 0 after all resets.

During a CALL instruction the stack pointer is first incremented, and the RAM location, pointed to by the STKPTR register is written with the contents of the PC (pointing to the instruction, following the CALL). During the RETURN instruction the contents of the RAM location pointed to by the STKPTR are transferred to the PC and then the stack pointer is decremented.

Status bits indicate if the stack is full, has overflowed or underflowed.

In our application the highest used level of stack was 5.

Below is shown the point in the software code, where the stack pointer is on the 3rd level

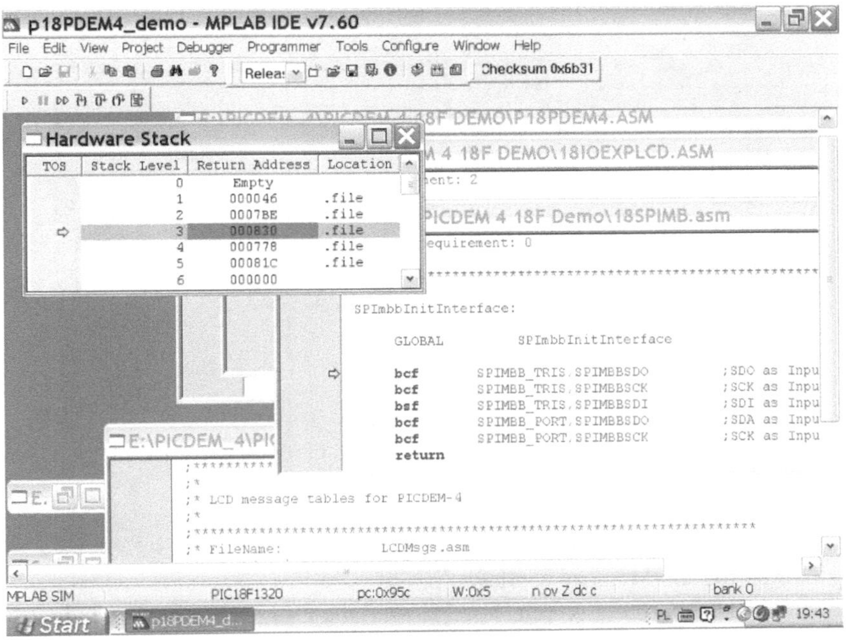

Fig.36. The hardware Stack

From the line # 34 (address 042) of the main code was called the function IoExpLCDInit (address 7ba). PC register was holding the address of the next instruction: 046

042 + 4 = 046

and pushed it on the first level of the stack. Program flow went to the location of 7ba, line # 990, which was the function: IoExpLCDInit. In the first line of this function was call to the function _InitIOExpander (line # 1047, address 82c. PC was holding the address of the next instruction:

7ba + 4 = 7be

and pushed this return value onto the stack as the 2nd level.

Then in the address 82c was executed macro mInitInterface, and from it was called the function SPImbbInitInterface. PC register was holding the address of the next instruction:

82c + 4 = 830

and again pushed this return value onto the stack (3rd level).

The green arrow on the left side points onto the top of the stack.

As we can see, there are some values on the levels 4 and 5, because after returning from function the program doesn't clear the stack location, but during subsequent pushing overwrites old values with the new.

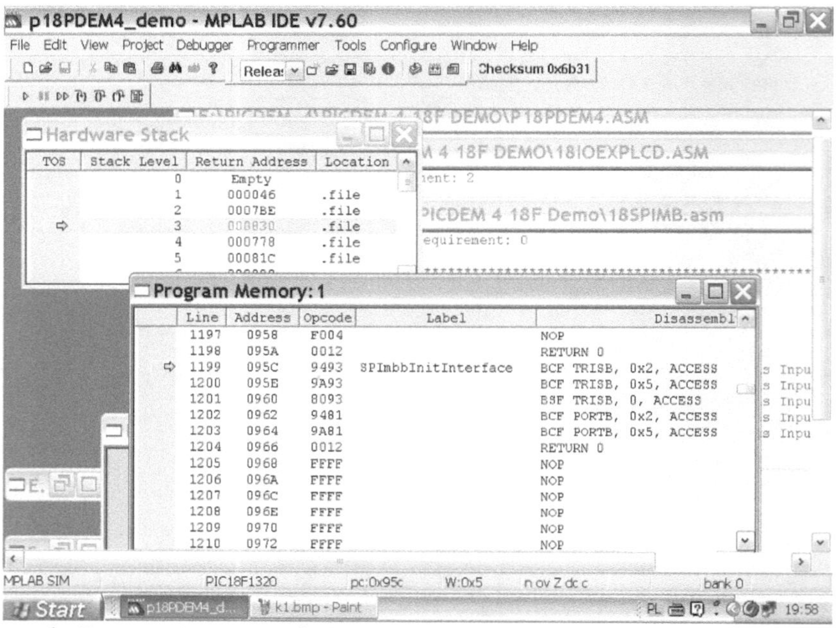

Fig.37. the Program Memory window

The second snapshoot screen shows as well the Program Memory window. The current line # is 1199 and the address of the function in the memory is 95c

We can present this as the following scheme:

the main code

…

line # 34

address 042

call to IoExpLCDInit

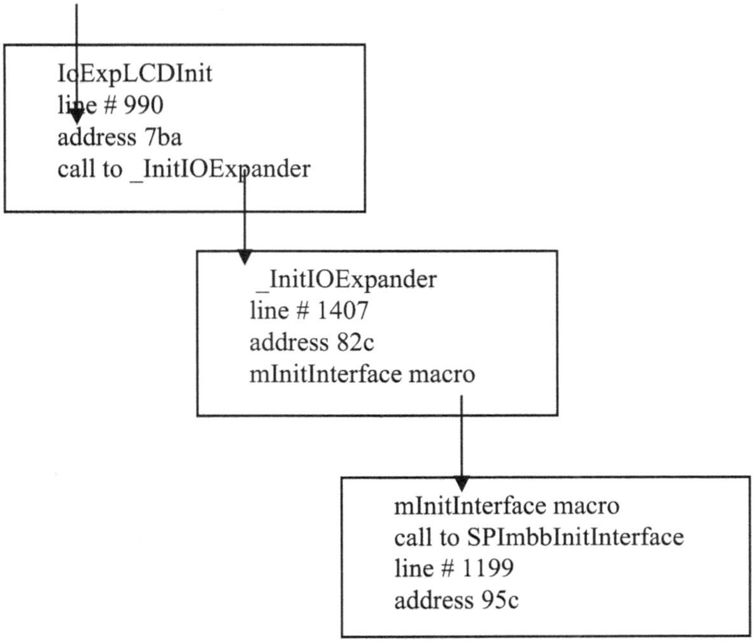

When we trace the program further on and execute return, then the stack pointer decrements and the green arrow on the left goes one level up.

Another information on, how the program works, we can get from the *.map* file. The *.map* file is generated, when we select this linker option in:

Project → Build Options → Project

and in the dialog window we open the linker tab.

Below is some part of the *.map* file.

```
MPLINK 4.11, Linker
Linker Map File - Created Wed Feb 27 12:18:47 2008

                                   Section Info
                  Section        Type     Address    Lo-
cation Size(Bytes)
                  ---------  ---------  ---------  ----
-----  ---------
              ResetVector          code      0x000000
program   0x000004
                   .cinit       romdata      0x00002a
program   0x000002
               Main_Start          code      0x00002c
program   0x00050a
```

```
                 IOExpLCDMsg          code       0x000536
program    0x0001ec
                IOExp2LCDCode          code       0x000722
program    0x00016e
                        PROG2          code       0x000890
program    0x00007a
                  CODE18SPIMBB          code       0x00090a
program    0x00005e
                      .config          code       0x300000
program    0x00000e
                        SPIMB         udata       0x000080
data     0x000017
            IOExpLCD_LCD_DATA         udata       0x000097
data     0x00000f
                      MainRAM         udata       0x0000a6
data     0x000009
                     MATH_VAR         udata       0x0000af
data     0x000009
                   MainOvrRAM         udata       0x0000b8
data     0x000003
                   MainAcsRAM         udata       0x0000bb
data     0x000001


                              Program Memory Usage
                               Start         End
                             ---------   ---------
                              0x000000    0x000003
                              0x00002a    0x000967
                              0x300000    0x30000d
               2384 out of 8472 program addresses used,
program memory utilization is 28%
```

Also the *.map* file provides information about all symbols in the output module. Symbols are: function names, labels for loop branches, the user defined register names etc. Information includes the address of the symbol, whether the symbol resides in program or data memory, whether it has external or static linkage and the name of the file, where it is defined, as below:

Symbols - Sorted by Address

```
                Name                          Address          Location
        Storage File
             ---------            ---------           ---------          --------- ----
-----
              start              0x00002c        program           static
        E:\. . . .mo\p18PDEM4.asm
            UserMode                         0x00007a        program
        static    E:\. . . .mo\p18PDEM4.asm
          VoltmeterMenu          0x00007a        program           static
        E:\. . . .mo\p18PDEM4.asm
         _Main_Start_007C        0x0000a8        program           static
        E:\. . . .mo\p18PDEM4.asm
         _Main_Start_0080        0x0000ac        program           static
        E:\. . . .mo\p18PDEM4.asm
         _Main_Start_00AE        0x0000da        program           static
        E:\. . . .mo\p18PDEM4.asm


  . . . . . . . . .

       _CODE18SPIMBB_004C   0x000956        program           static
        E:\. . . .mo\18SPIMB.asm
       SPImbbInitInterface        0x00095c        program           extern
        E:\. . . .mo\18SPIMB.asm
           SPImbbBuffer          0x000080        data              extern
        E:\. . . .mo\spimb.asm
           SPImbbRxData          0x000094        data              static
        E:\. . . .mo\spimb.asm
            SPImbbData           0x000095        data              extern
        E:\. . . .mo\spimb.asm
          SPImbbDataRead         0x000096        data              static
        E:\. . . .mo\spimb.asm
             _ioExpDelay                     0x000097        data
        static    E:\. . . .mo\IoExpLcd.asm
      _ioExpInternalCounter       0x000098        data              static
        E:\. . . .mo\IoExpLcd.asm
         _ioExpDataToLCD         0x000099        data              static
        E:\. . . .mo\IoExpLcd.asm
       _ioExpCommandToLCD              0x00009a        data
        static    E:\. . . .mo\IoExpLcd.asm

  . . . . . . . . .
```

<u>The lister file .lst</u>

One of the output file generated by the linker is the text file *.lst*, called lister file. Inspecting this file allows to understand, how the program works. Let us see the beginning of the *.lst* file of the demo program p18PDEM.asm:

```
;*************************************************
  ;Reset                                      Vector
;************************************************
  ResetVector        CODE     0x00

  000000    ef16       GOTO        0x2c
                     goto   start
  000002    f000


  Main_Start         CODE

   start
   ;Initialize I/O pins
  00002c    0e70       MOVLW       0x70
                     movlw                        B'01110000'
                ;Select 8MHz frequency
  00002e    6ed3       MOVWF       0xd3,0x0
                     movwf    OSCCON
  000030    6a80       CLRF        0x80,0x0
                     clrf     PORTA
  000032    6a81       CLRF        0x81,0x0
                     clrf     PORTB
  000034    68c1       SETF        0xc1,0x0
                     setf     ADCON1

  000036    0e3f       MOVLW       0x3f
                     movlw    B'00111111'
  000038    6e92       MOVWF       0x92,0x0
                     movwf    TRISA
  00003a    0ef7       MOVLW       0xf7
                     movlw    B'11110111'
  00003c    6e93       MOVWF       0x93,0x0
                     movwf    TRISB
```

```
00003e    ec1f        CALL        0x43e,0x0
                    call    LongDelay
                                        ;Provide  some  delay
                                  to match-up reset time
000040    f002
                                        ;of IO Expander
```

First four bytes of the reset vector is a reset section.

From the address of 0x020 on goes the Main-Start code, comprising 0x00050a bytes of code. The first line of the Main-Start coOde is the line:

```
00002c    0e70        MOVLW       0x70
                    movlw                           B'01110000'
              ;Select 8MHz frequency
```

2ch is the address of the line in hex.

Oe is the op code name of the instruction *movlw*

70 is the value of the argument, equal 0x70. This value is to be loaded into WREG register.

The next line with the address of 00002e is:

```
00002e    6ed3        MOVWF       0xd3,0x0
                    movwf    OSCCON
```

6e is the op code of the instruction *movwf*

d3 is the address of the SFR OSCCON register

The result of running this instruction can be watched in the window of MPLAB IDE:

View → Special Function Registers, and also in the window:

View → View → File Registers

as is shown below:

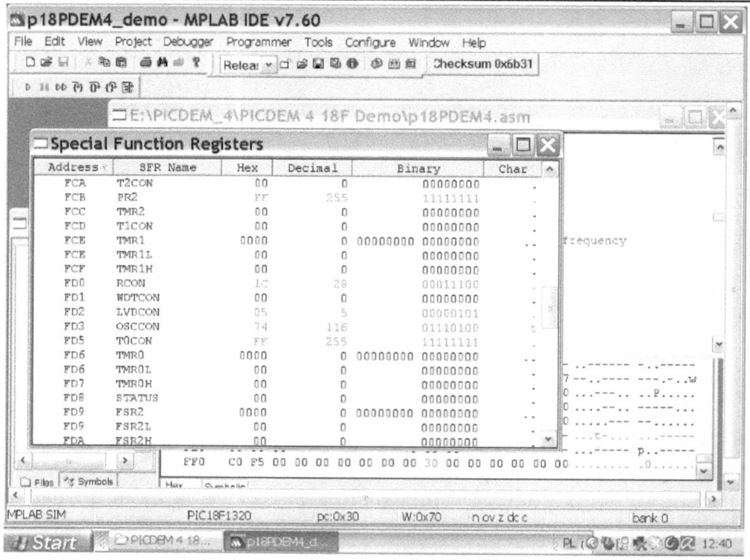

Fig.38. The result of running this instruction watched in the window

Let us remember, that OSCCON register has the address of Fd3.

As was stated, the Special Function Registers we can also inspect in the window with all data memory, named: File Registers. We have to scroll down the window to the very end to see the upper addresses of the memory. There are no names of registers in this window, but we remember, that OSCCON register has the address of FD3

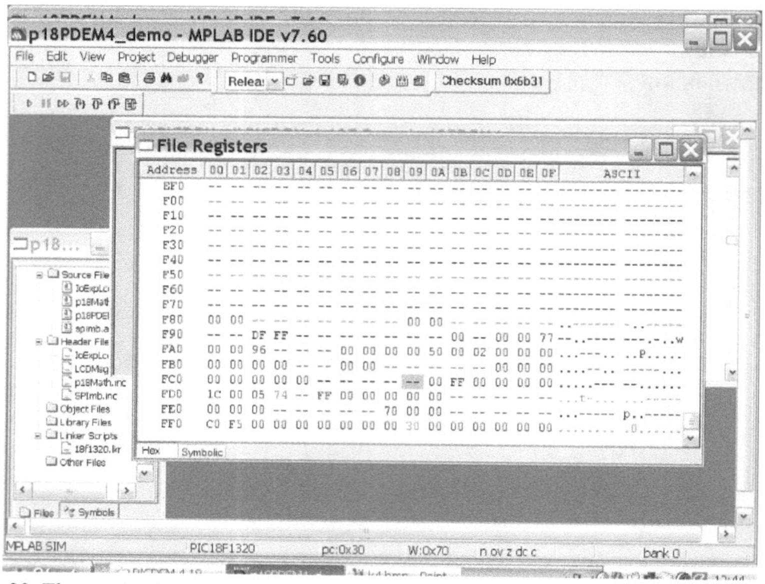

Fig.39. The result of running this instruction watched in the window

As we see, the working register WREG (address 0xFE8) holds the value of 70, while the OSCCON register has the value of 74. The third bit in the register is set, because the INTOSC frequency is stable

Let us inspect the tenth and eleventh lines of code:

```
00003e    ec1f      CALL       0x43e,0x0
                 call     LongDelay
                                   ;Provide  some  delay
                            to match-up reset time
000040    f002
                                   ;of IO Expander
```

0x00003e is the instruction address in the program memory.

ec is the op code of the *call* instruction.

Let us leave the value of 1f as it is, and take into account the argument of the instruction: 0x43e. It points to the address in the program memory, where resides the function: *LongDelay*.

0x0 denotes the memory stack, where the return address is pushed (PC + 4).

Tables, Table instructions

As we can see in the *.map* file:

```
MPLINK 4.11, Linker
Linker Map File - Created Wed Feb 27 12:18:47 2008

                                    Section Info
                      Section       Type     Address    Lo-
cation Size(Bytes)
                    ---------   ---------  ---------   ----
-----  ---------
                  ResetVector          code       0x000000
program    0x000004
                       .cinit      romdata        0x00002a
program    0x000002
                   Main_Start          code       0x00002c
program    0x00050a
                  IOExpLCDMsg          code       0x000536
program    0x0001ec
```

```
                IOExp2LCDCode            code      0x000722
program    0x00016e
```

the IOExpLCDMsg code section for displaying data on LCD resides in the program memory, starting with the address of 0x000536.

The first string to display is:

"____Microchip___"

and is labeled as MsgMicrochip. The string is expanded as follows:

```
  IOExpLCDMsg      code

   IOExpLCDStringTable:                  ;table for standard
messages

               GLOBAL  IOExpLCDStringTable

    MsgMicrochip:
  000536    2020     ADDWFC     0x20,0x0,0x0      data  "
Microchip    ",0    ;
  000538    2020     ADDWFC     0x20,0x0,0x0
  00053a    694d     SETF       0x4d,0x1
  00053c    7263     BTG        0x63,0x1,0x0
  00053e    636f     CPFSEQ     0x6f,0x1
  000540    6968     SETF       0x68,0x1
  000542    2070     ADDWFC     0x70,0x0,0x0
  000544    2020     ADDWFC     0x20,0x0,0x0
  000546    0000     NOP
```

The first column contains addresses of particular words of code. The second column contains ASCII characters to be displayed, the lower byte first.:

2020	space, space	
2020	"	"
694d	i	M
7263	r	c
636f	c	o
6968	i	h
2070	_	p
2020	_	_

"_" also denotes space.

The third and fourth columns can be neglected, because are no relevant.

In the 18th line of the Main-code section the following macro is invoked:

```
;Display "Microchip"
00004e    0e00      MOVLW       0x0
                  mIOExpLCDSendMessage     MsgMicrochip
000050    6ef8      MOVWF       0xf8,0x0
000052    0e05      MOVLW       0x5
000054    6ef7      MOVWF       0xf7,0x0
000056    0e36      MOVLW       0x36
000058    6ef6      MOVWF       0xf6,0x0
00005a    ec91      CALL        0x722,0x0
00005c    f003
```

Notes, that 0xF8, 0xF7 and 0xF6 are addresses of three registers of the TBLPTR: upper, high and low.

The macro gets as the argument: MsgMicrochip, which is the label in the first string in the table.The pointer to the symbolic name MsgMicrochip is 0x000536. This address has to be moved to the TBLPTR register, to be read by the TBLRD (Table Read) instruction.

The TBLPTR register is comprised of three SFR registers:

Table Pointer Upper byte
Table Pointer High byte
Table Pointer Low byte
(TBLPTRU:TBLPTRH:TBLPTRL).

The macro mIOExpLCDSendMessage divides the pointer to MsgMessage into three parts and moves them to above mentioned Table Pointer bytes. Those three parts are:

00	to	TBLPTRU
05	to	TBLPTRH
36	to	TBLPTRL.

Then the macro invokes the function:

```
00005a    ec91      CALL        0x722,0x0
```

(call IOExpLCDSendMessage).

The address of the function is 000722 in hex. and the respective code with the line addresses is shown below:

```
IOExpLCDSendMessage:

                  GLOBAL            IOExpLCDSendMessage

                   IOExpLCDNextMessageBit
000722    0009      TBLRDPOSTINC
```

```
                TBLRD*+
          ;read from table
000724    50f5     MOVF        0xf5,0x0,0x0
                movf                            TABLAT,w
          ;if 0 end
000726    0900     IORLW       0x0
                iorlw            0x00
000728    b4d8     BTFSC       0xd8,0x2,0x0
                btfsc            STATUS,Z
00072a    0012     RETURN      0x0
                return
00072c    ecbd     CALL        0x77a,0x0
                call                   IOExpLCDWriteData
          ;otherwise write character in LCD
00072e    f003
000730    ef91     GOTO        0x722
                goto             IOExpLCDNextMessageBit
000732    f003
```

The table is read with the instruction postincremented TBLRD*+. Then the content of the TBLAT register is copied to the W register. TBLAT register holds the value (ASCII character), pointed to by the TBLPTR register.

The contents of the W register is checked (iored with the value of 0x0), if it is the end of the string. If not yet, then the function writing data is invoked and the next character is to be loaded to the WREG register in the loop.

If the end of the string is encounterwed, the Return is executed.

The results of executing the code can be inspected by watching File Registers and Special Function Registers Memory, especially TBLPTRU, TBLPTRH, TBLPTRL and TBLAT registers

Data memory

Now we describe briefly the data memory. The data memory map is shown below, in Figure 40

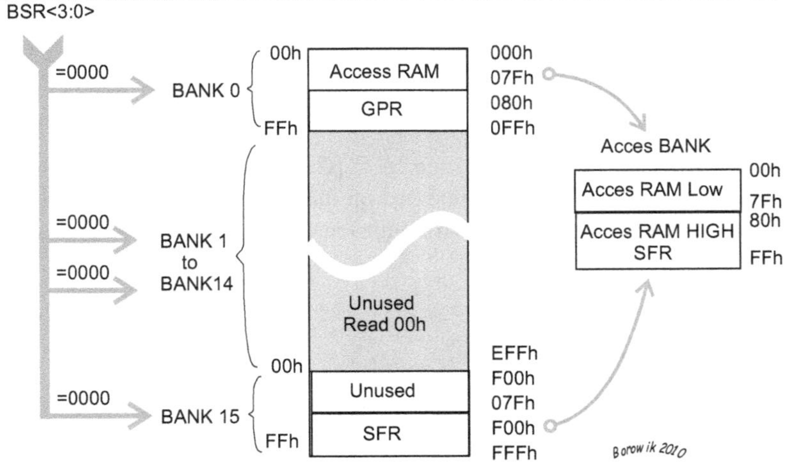

Fig.40 Program memory map and stack for PIC18F1320

The lowest address of 00h is in the top. Addresses grow up when we go down the figure. At the very top (bottom on the figure) are located SFR. They occupy addresses EFFh to FFFh (256 bytes).

Lower part of data memory occupy GPR (General Purpose Registers). These are user defined registers, used for data storage and scratch pad operations in the user application.

As already has been described, we can view data memory selecting in the MPLAB IDE:

View → Special Function Registers,

and to see GPR:

View → File Registers

Then we can scroll down the screen and see the SFR registers as well (see fig. 40).

The application of the PIC24FJ microcontroller with the 240x128 LCD display and the analog accelerometer sensor.

The source code has been written in C (C30 compiler). The sensor used is Freescale MMA7260QT, accommodated on the sensor board of Sure Electronics. For the convenience we put the board with sensor to the metal box as shown in the figure.

Fig.41 The accelerometer MMA7260QT, accommodated on the sensor board

Schematic diagram of the accelerometer with operational amplifiers is shown in figure 42.

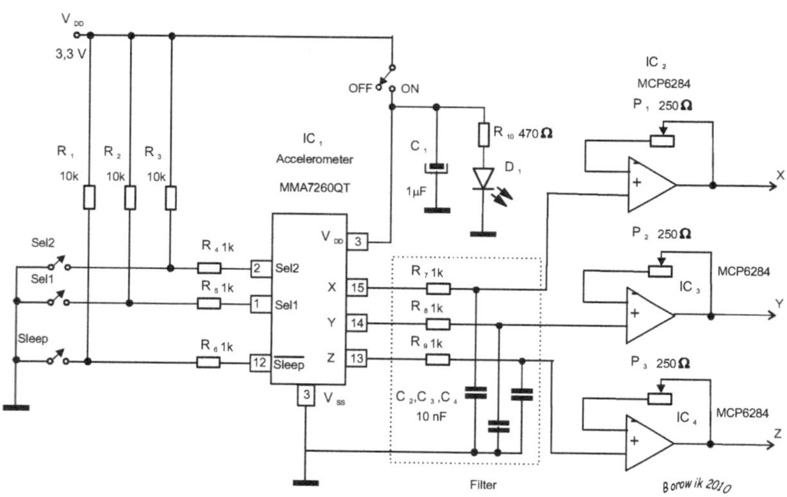

Fig.42 The accelerometer MMA7260QT, schematic diagram

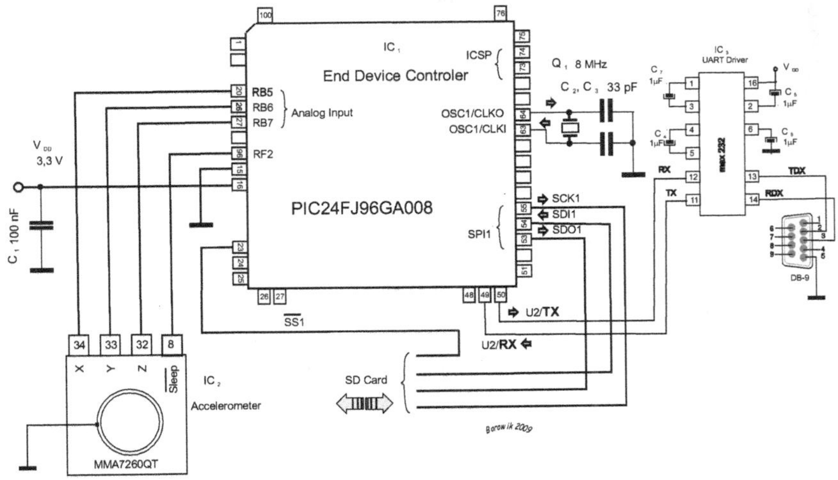

Fig.43 The accelerometer application circuit.

After powering up, the program waits for pressing the switch (connected to RC3). Then on the LCD display appears the text: "a_z = ". Also on the left side of the screen appear in the first column the marks "-". They allow to imagine, where is the Mid of the vertical scale.

Now system waits for data from accelerometer attached to the cable. When we shake the box with accelerometer, then the processor calculates if data from sensor are greater than sensor noise. If so, then 22 results are collected.

Reason of the number 22 is, that in the application LCD operates in the text mode. It has 30 columns and 16 rows. Last 7 columns are reserved for numerical values of the data, while the first column has scale marks. Therefore only 30 – 7 – 1 = 22 results are needed for scattergram.

First the data are collected and saved in the array. After finishing the sequence the scattergram is plotted in the text mode and first numeric values are presented on the screen, as shown in figures a, b and c.

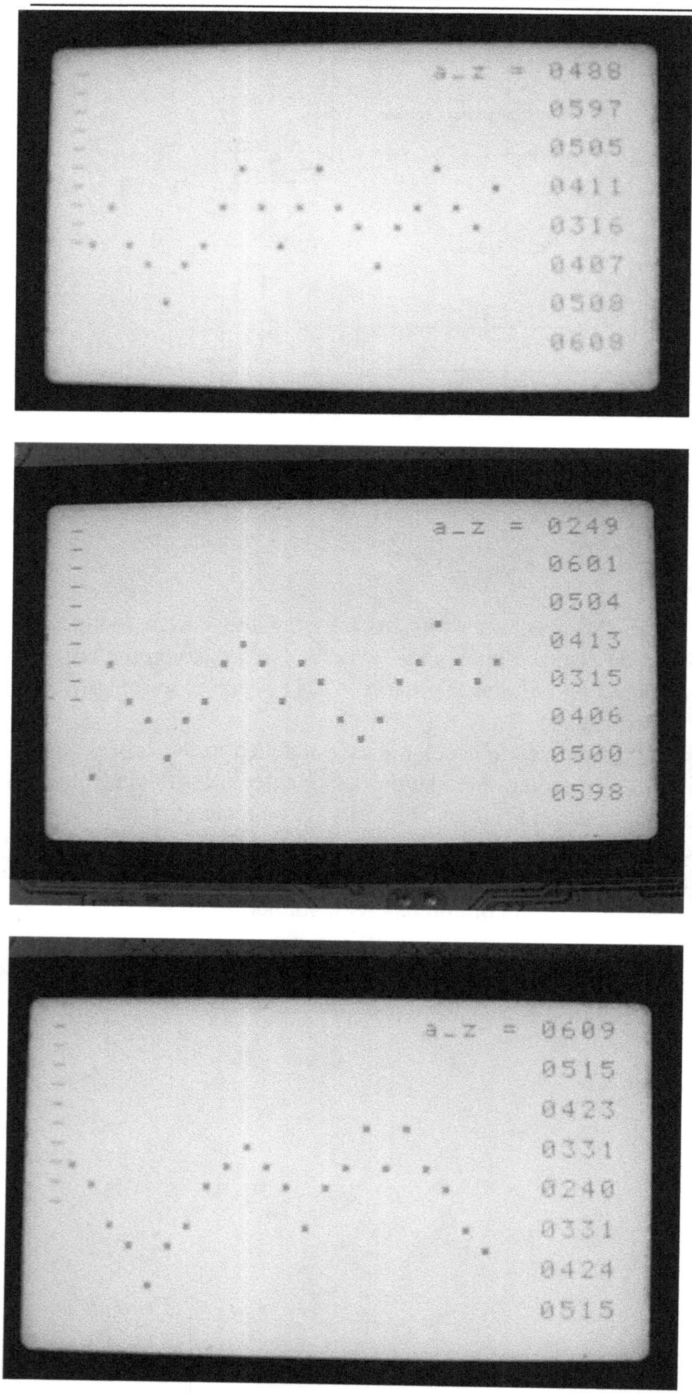

Fig.44 Scattergram with data from accelerometter

Those values are raw data not converted to acceleration units, nor to mVolts yet. For 10-bits ADC module they are in the range of 0 to 1024. The center of the scattergram is about the value of 500. For better clarity we truncated lowest and highest part of the scale.

Application runs according to following two flowcharts:

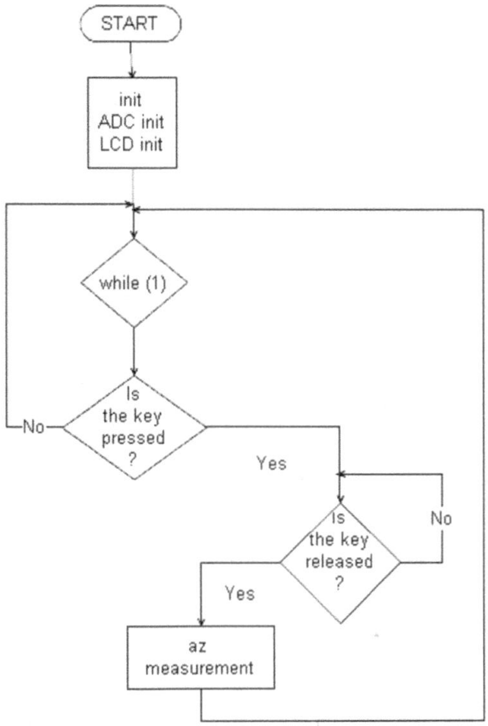

Fig.45 Aplication flowchart nr 1

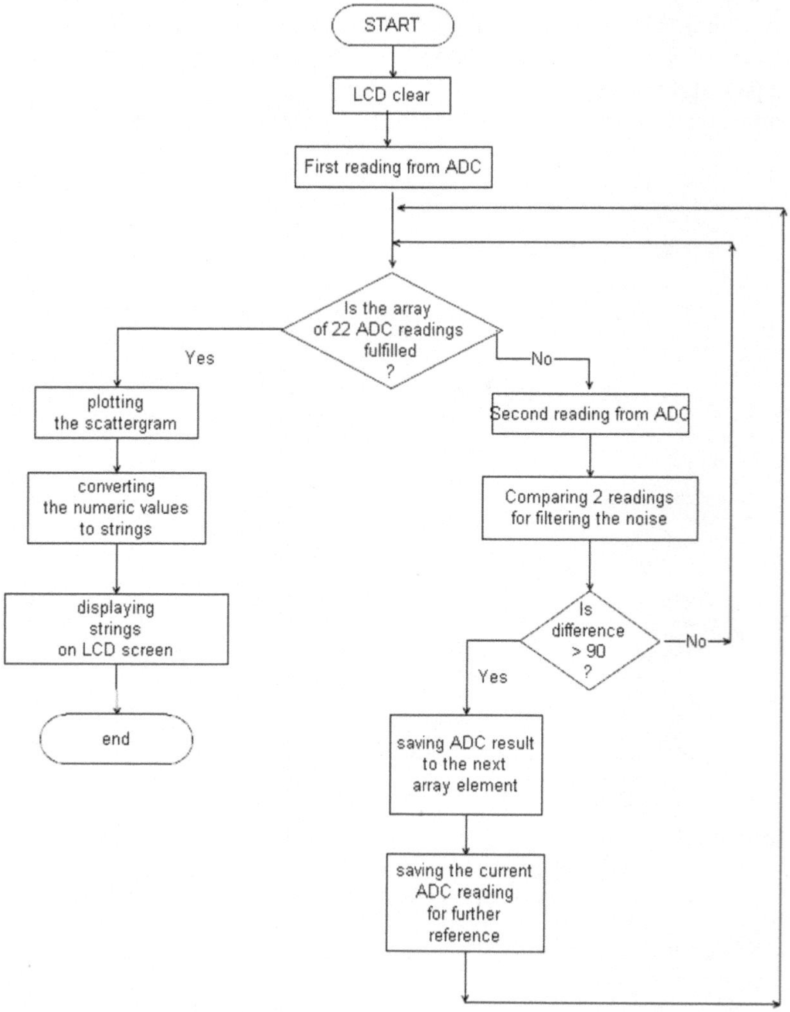

Fig.46 Aplication flowchart nr 2

Listing of the source code is shown below. Detailed description is presented after that.

```
//---------------------

#include<p24fxxxx.h>
#include  <stdlib.h>
#include <math.h>
```

```
      _CONFIG2( FNOSC_FRCDIV ) //
      _CONFIG1(JTAGEN_OFF & FWDTEN_OFF)    // JTAG off,
watchdog timer off

  #define  wr     LATFbits.LATF1    //  WR  L Write
  #define  rd     LATFbits.LATF0    //  RD  L READ
  #define  cs     LATFbits.LATF3    //   CS   L Chip Se-
lect
  #define  cd     LATFbits.LATF2    //   CD   L DATA    H
Control
  #define  rst    LATFbits.LATF5    //  RST  L Reset
  #define  fs     LATFbits.LATF4    //  FS  FONT SELECT
  // polish national characters
  char pol [] = {
  0x00, 0x0e, 0x01, 0x0f, 0x11, 0x0f, 0x02, 0x04,    //
a
  0x02, 0x04, 0x0e, 0x11, 0x10, 0x11, 0x0e, 0x00,    //
c
  0x00, 0x0e, 0x11, 0x1f, 0x10, 0x0e, 0x02, 0x04,    //
e
  0x02, 0x04, 0x0f, 0x10, 0x1e, 0x01, 0x1f, 0x00,    //
s
  0x0c, 0x04, 0x06, 0x04, 0x0c, 0x04, 0x0e, 0x00,    //
l
  0x02, 0x04, 0x0e, 0x11, 0x11, 0x11, 0x0e, 0x00,    //
o
  0x04, 0x00, 0x1f, 0x02, 0x04, 0x08, 0x1f, 0x00,    //
z
  0x02, 0x04, 0x1f, 0x02, 0x04, 0x08, 0x1f, 0x00,    //
z
  0x02, 0x04, 0x16, 0x19, 0x11, 0x11, 0x11, 0x00,    //
n
  0x10, 0x10, 0x18, 0x10, 0x30, 0x10, 0x1f, 0x00,    //
L
  0x00, 0x00, 0x00, 0x1c, 0x1c, 0x1c, 0x00, 0x00     //
kropka
  };

  char t1[4];
  char t2[5];

  // #define ay1       0            // 10k ay connected
to RB0
```

```
  // #define ax1        1          // 10k ax connected
to RB1
  #define az1        5          // 10k az connected
to RB5
  #define AINPUTS 0xff00     // Analog input for sen-
sor on line 5


  void init();
  void initADC( int amask);
  void lcd_init();
  void delay();
  void delay2();
  void delay3();
  void txt_hm();
  void send_d(unsigned int x);
  void send2_d(unsigned int x);
  void send_i(unsigned char x);
  void chk_busy();
  void chk_busy2();
  void gr_hm();
  void os_x(unsigned int ay);
  void ekran0();
  unsigned int readADC( int ch);
  void os_y(unsigned int ax);
  void g_setdot(unsigned int a1, unsigned int a2);
  void lcd_cl();
  void put_char(char a, unsigned int a1, unsigned int
a2);
  void putsLCD( char *s, unsigned int a1, unsigned int
a2);
  void gr_clear();
  void  num2str(unsigned int a);

  int main(void)
  {
      initADC( AINPUTS);  // initialize the ADC
      delay2();

      init();
      lcd_init();

      while (1)
      {
            if(PORTCbits.RC3)
```

```
                {
                        while (PORTCbits.RC3);
                        ekran0 ();
                }
    }
    return 0;
}


// ------------------------------
void ekran0()
{
    int i, j;
    unsigned int b, a, a_max, a_min;
    unsigned int az[30];
    send_i(0x94);               //text  on
    lcd_cl();
    for (i = 0; i < 10; i++)
        put_char('-', 0, i);

    a_max = 0;
    a_min = 500;
    putsLCD("a_z = ", 19, 0);

    j = 0;
    a =  readADC(az1);

    // collecting 22 measurements,
    // filtering the noise, when adjacent data differ
    // less, than 90 mV

    while (j < 22)
    {
        b =  readADC(az1);
        i = a - b;
        if (i < 0)
            i *= -1;
        delay();
        if (i > 90)
        {    az[j] = a;
            delay();
            j++;
            a = b;
        }
```

```
            delay();
      }

      // displaying 8 results as numbers
      for (i = 0; i < 9; i++)
      {
            num2str(az[i]); //convert num to string t1
            putsLCD(t1, 25,i * 2);
      }

      // making plot diagram of 22 points
      for (i = 0; i <22; i++)
      {
            j = az[i]/64;
            put_char(0xaa, i + 1, 16 - j);
      }
}

//---------------------------------------------
void init()
{
     TRISC=0X001A;
     TRISG=0X0000;
     TRISB=0X00ff;
     TRISD=0X0000;
           TRISF=0X0000;
     SPI2CON1=0x0278;
     SPI2STAT=0x8000;
    }

// initialize the ADC for single conversion, select
// Analog input pins
void initADC( int amask)
{
     AD1PCFG = amask;    // select analog input pins
     AD1CON1 = 0x00E0;   // auto convert after end of
                              // sampling
     AD1CSSL = 0;        // no scanning required
     AD1CON3 = 0x1F02;   // max sample time = 31Tad,
                       //  Tad = 2 x Tcy = 125ns >75ns
     AD1CON2 = 0;          // use MUXA, AVss and Avdd
                       // are used as Vref+/-
     AD1CON1bits.ADON = 1; // turn on the ADC
} //initADC
```

```
//---------------------------
void lcd_init()
{
     int i;

     wr = 1;
     rd = 1 ;
     cs = 1;
     cd = 1;
     fs = 0 ;
     rst= 1;
     delay2();
     rst = 0;
     delay();
     rst = 1;

     delay2();
     txt_hm();
     gr_hm();
     send_i(0x80);   // OR mode
     send_i(0x98);             // graph on


     gr_clear();

     // generating user defined characters

     send_d(0x04);     // offset = 4
     send_d(0x00);
     send_i(0x22);             // set offset register

     send_d(0x00);
     send_d(0x24);             // if offset = 4, then CG
RAM starts at 0x2400
     send_i(0x24);

     send_i(0xb0);             // data auto write

     for (i=0; i<88; i++)  //a, c, e, s, l, o, z, z, n
     {     send2_d(pol[i]);
     }
     send_i(0xb2);     // auto reset
}
```

```
//------------------------------------------
void delay()
{
    int m;
    for(m=0;m<10;m++)
       {;}
}


//------------------------------------------

void delay2()
 {
   unsigned int i2;
   for(i2=0;i2<500;i2++);
 }

//------------------------------------------

void delay3()
 {
   unsigned int i3;
   for(i3=0;i3<2000;i3++)
          delay2();
 }


// - - - - - - - - - - - - - -
unsigned int readADC( int ch)
{
    AD1CHS  = ch;                 // select analog
                    //input channel
    AD1CON1bits.SAMP = 1;         // start sampling,
              //  automatic conversion will follow
    while (!AD1CON1bits.DONE);     // wait to complete
                           // the conversion
    return (unsigned int) ADC1BUF0;  // read the
                           // conversion result
} // readADC

// -------------------------------
void os_y(unsigned int ax)
```

```
{
    int i;
    gr_hm();
    for (i=0; i<128; i++)
          g_setdot(ax,i);
}

// -------------------------------

void os_x(unsigned int ay)
{
    int i;
    gr_hm();
    for (i=0; i<240; i++)
          g_setdot(i,ay);
}

// -------------------------------
void g_setdot(unsigned int a1, unsigned int a2)
{
// a1 - coord. x are 0 to 239
// a2 - coord. y are 0 to 127

    unsigned int adr, cmd;

    adr = a1/8 + 30 * a2;

    // 128 rows x 30 bytes in each row

    cmd = 0xf8 |  (8 - (a1 % 8));       // bitwise OR
                 // determine, which bit is to be set

    gr_hm();

    send_d(adr % 256);          // younger byte
    send_d(0x10 + adr/ 256); // more significant byte
    // gr. home address + byte no in the RAM memory

    send_i(0x24);

    send_i(cmd);
    }
```

```
//-------------------------------------------
void txt_hm()  //Text Home
{
    send_d(0x00);
    send_d(0x00);
    send_i(0x40);           // Set Text Home Address

    send_d(0x1e);           // 1e = 30 columns
    send_d(0x00);
    send_i(0x41);       // Set Text Area
}

//-------------------------------------------
void send_d(unsigned int x)
{
  chk_busy();
    cd = 0;
    cs = 0;
    wr = 0;
    PORTD = x;
    delay();
    delay();
    wr = 1;
    cs = 1;
}

//-------------------------------------------
void send2_d(unsigned int x)      // for AUTO mode
          // with different function chk_busy( )
{
  chk_busy2();
    cd = 0;
    cs = 0;
    wr = 0;
    PORTD = x;
    delay();
    delay();
    wr = 1;
    cs = 1;
}

//-----------------------------
void send_i(unsigned char x)
{
    chk_busy();
```

```
    cs = 0;
    wr = 0;
    PORTD = x;
    delay();
    delay();
    wr = 1;
    cs = 1;
}

//-------------------------------
void chk_busy()
{
    cd = 1;
    wr = 1;
    TRISD=0Xffff;
    do {
          cs = 1;
          rd = 1;
          delay();
          cs = 0 ;
          rd = 0 ;
          }
    while(!PORTDbits.RD0 | !PORTDbits.RD1 );
    cs = 1 ;
    rd = 1 ;
    TRISD=0X0000;
 }

//------------------------------------------
void chk_busy2()
{
    cd = 1;
    wr = 1;
    TRISD=0Xffff;
    do {
          cs = 1;
          rd = 1;
          delay();
          cs = 0 ;
          rd = 0 ;
          }
// checking Auto mode data write capability (bit 3)

     while(!PORTDbits.RD3 );
```

```
    cs = 1 ;
    rd = 1 ;
    TRISD=0X0000;
 }

//-------------------------------------------
void gr_hm()
{
    send_d(0x00);
    send_d(0x10);            // 0x1000
    send_i(0x42);     // Set Graphics Home Address

    send_d(0x1e);            // 30 (x 8)
    send_d(0x00);
    send_i(0x43);       // Set Graphics Area
}


//----------------
void lcd_cl()
{
int i, j;
for(i=0;i<16;i++)
    for(j=0;j<30;j++)
         put_char(' ', j, i);
}


// -------------------------------
void put_char(char a, unsigned int a1, unsigned int
a2)
{
    unsigned int adr;
    adr = 30 * a2 + a1;
    txt_hm();

    send_i(0x94);     // Text on

    send_d(adr % 256);
    send_d(adr / 256);
    send_i(0x24);     //   Set Address Pointer

    send_d(a - 32);   //  ASCII offset
```

```
    send_i(0xc0);     //Data Write, increment address

    send_i(0x9c);     // Text and Graph on
  }


  // -------------------------------
  void putsLCD( char *s, unsigned int a1, unsigned int
a2)
  {
      int i = 0;
      while( *s)
      {     put_char( *s++, a1 + i, a2);
            i++;
            delay();
      }
  } //putsLCD


  //---------------
  void gr_clear()
  {
      int i, j;
      send_d(0x00);
      send_d(0x10);
      send_i(0x24);

      send_i(0x98);           // text off, graphic on

      for(i=0;i<30;i++)
            for(j=0;j<128;j++)
            {
            send_d(0x00);
            send_i(0xc0); //data write, increment addr.
            }
  }

  // -------------------------------

  void  num2str(unsigned int b)
  {
  int i, j, bas1;
  if (b > 1024)
      {t1[0] = 'e';     // error, when value > 1024
```

```
      t1[1] = '\0';
      return ;
      }
  for (i=0, j=4; i < 4; i++, j--)
      {
              bas1 = pow(10, j - 1);  // 1000, 100, 10, 1
              t1[i] = (b / bas1) + 48;
              b %= bas1;
      }
  t1[i] = '\0';
  }

  // ------------------------------
```

The program occupies 4% of the program memory, as shown below:

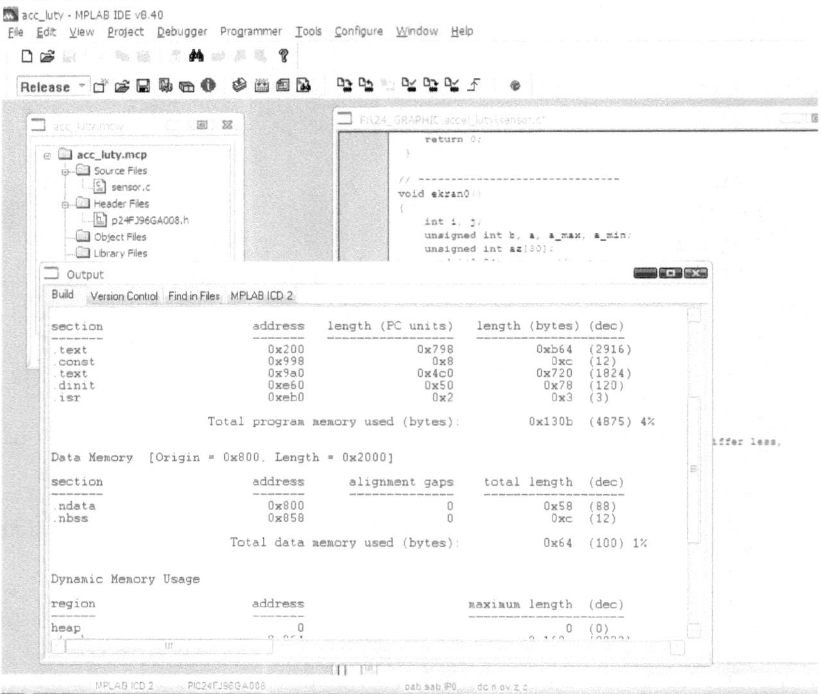

Fig.47 Ocupation of program and data memory

The special part of the program is devoted to communicating with LCD module. The detailed description follows.

Interfacing microcontroller to LCD display

The LCD module has LCD screen 240 x 128 driven with T6963C controller, operating in text and/or graphic mode.

The initial settings of the controller are:

Pin name	state
~DUAL	H
MDS	H
MD1	L
MD0	L

// 16 lines, 128 vertical dots, Single Screen

MD2	L
MD3	H
FS0	L
FS1	L

// 40 columns, 8x8 font.
F osc 4MHz.

Number of columns is adjusted in the software and equals 30 in the text mode.
If we set the number of columns to 30, then addresses of particular row in the graphic area are:

1 line addresses:	0x00	to	0x1d
2	0x1E ...		
3	0x3C		
4	0x5A		
5	0x78		
6	0x96		
7	0xB4		
8	0xD2		
9	0xF0		
10	0x10E		
11	0x12C		
12	0x14A		
13	0x168		
14	0x186		
15	0x1A4		
16	0x1C2 etc.		

The controller has its set of commands. Most important of them are:

$21 Cursor Pointer set
$24 Address Pointer Set

$40 Text Home address set
$41 Text Area Set
$42 Graphics Home Address Set
$43 Graphics Area Set

$80 OR mode command, Internal CG ROM Mode
$81 XOR mode
$83 AND mode
$84 Text Attribute Mode
$88 External CG ROM Mode

$90 Display off
$91 Blink on
$92 Cursor on
$94 Text on
$98 Graphic on
$9C Text and graphic on

// Cursor pattern
$A0 Cursor 1 line
$A1 Cursor 2 line
$A2 Cursor 3 line
$A3 Cursor 4 line
$A4 Cursor 5 line
$A5 Cursor 6 line
$A6 Cursor 7 line
$A7 Cursor 8 line

$B0 Data Auto Write
$B1 Data Auto Read
$B2 Auto Data Reset

$C0 Data write and increment address
$C1 Data read and increment address
$C2 Data write and decrement address
$C3 Data read and decrement address
$C4 Data write, no address change
$C5 Data read, no address change

// Bit Set/Reset (OR with bit number 0-7)

$F0 Bit reset
$F8 Bit set

Most commands require arguments (data). In the software first we set data with the function send_d( ), then the command/instruction with the function send_I( ). If there are 2 bytes of data, less significant byte goes first, then more significant byte.

Example
Graphic home address is 0x1000.
We send first 0x00, then 0x10.

Now let's consider the listing of the application.

LCD controller T6963C requires 6 control lines and 8 data lines to communicate with the microcontroller. The control lines are designated as follows:

```
#define  wr     LATFbits.LATF1    //  WR  L Write
#define  rd     LATFbits.LATF0    //  RD  L READ
#define  cs     LATFbits.LATF3    // CS L Chip Select
#define  cd     LATFbits.LATF2   //CD L DATA H Control
#define  rst    LATFbits.LATF5    //  RST  L Reset
#define  fs     LATFbits.LATF4    //  FS  FONT SELECT
```

All of them are connected to port F.
Data is send through lower part of port D.

Both ports are initially set to be output ports:

```
TRISD=0X0000;
    TRISF=0X0000;
```

```
void init()
{
        . . . . .
    TRISD=0X0000;
          TRISF=0X0000;
    . . . . .
   }
```

We initialize LCD with the function lcd_init( ).

```
void lcd_init()
{
    int i;
          // initial settings for control lines
    wr = 1;
    rd = 1 ;
    cs = 1;
    cd = 1;
    fs = 0 ;
    rst= 1;
    delay2();
    rst = 0;
    delay();
    rst = 1;

    delay2();
    txt_hm();    // to set text home address
                 // in the memory area
// and the number of columns in the text mode

    gr_hm();     // to set graphic home address
                 // and the number of columns
                       // in the graphics mode

    send_i(0x80);      // OR mode
    send_i(0x98);      // graph on


    gr_clear();


    // generating national polish characters

    send_d(0x04);      // offset  4
```

```
    send_d(0x00);
    send_i(0x22);                // set offset register

    send_d(0x00);
    send_d(0x24);          //CG RAM address starting
                           //at 0x2400 for offset equal 4
    send_i(0x24);

    send_i(0xb0);                // data auto write

          // sending user defined characters
                // to CG RAM memory

    for (i=0; i<88; i++)  //a, c, e, s, l, o, z, z, n
    {     send2_d(pol[i]);
    }
    send_i(0xb2);          // auto reset
}
```

The text data, graphics data and external CG data can be freely allocated to the display memory area (64 KB max). We set home address to 0x0000 in text mode and for graphics: 0x1000.

```
void txt_hm()  //Text Home
{
    send_d(0x00);
    send_d(0x00);
    send_i(0x40);                // Set Text Home Address

    send_d(0x1e);                // 1e = 30 kolumn
    send_d(0x00);
    send_i(0x41);            // Set Text Area
}

//---------------------------------------------

void gr_hm()
{
    send_d(0x00);
    send_d(0x10);                // 0x1000
    send_i(0x42);          // Set Graphics Home Address
```

```
    send_d(0x1e);      //30 (x 8 bits for each column)
    send_d(0x00);
    send_i(0x43);        // Set Graphics Area
}
```

The character codes used by the T6963 are different from ASCII codes. The user must subtract 0x20 (decimal 32) from the ASCII code sending it to the display, as is in the function put_char( ).

```
void put_char(char a,unsigned int a1,unsigned int a2)
{
    unsigned int adr;
    adr = 30 * a2 + a1;
    txt_hm();

    send_i(0x94);      // Text on

    send_d(adr % 256);
    send_d(adr / 256);
    send_i(0x24);      //   Set Address Pointer

    send_d(a - 32);    //  ASCII offset

    send_i(0xc0);      // Data Write,increment address

    send_i(0x9c);      // Text and Graph on
}

// -------------------------------
```

The last function described is num2str( ). It converts 4 digit numeric value to text string.

```
void  num2str(unsigned int b)
{
int i, j, bas1;
if (b > 1024)            // error, when value > 1024
    {t1[0] = 'e';
    t1[1] = '\0';
    return ;
    }
```

```
for (i=0, j=4; i < 4; i++, j--)
    {
          bas1 = pow(10, j - 1);  // 1000, 100, 10, 1
          t1[i] = (b / bas1) + 48;
          b %= bas1;
    }
t1[i] = '\0';   // string terminator
}
```

References

1. www.microchip.com;
2. Tomasz Jabłoński, *Mikrokontrolery PIC16F8x w praktyce*, BTC, Warszawa 2002
3. Stanisław Pietraszek, *Mikroprocesory jednoukładowe* PIC, Helion, Gliwice 2002
4. PIC16F8X Data Sheet, DS30430B, Microchip Technology Inc., 1996;
5. PIC18FXX2 Data Sheet, DS39564B, Microchip Technology Inc., 2002;
6. PIC18F452 to PIC18F4520 Migration, DS39647A, Microchip Technology Inc., 2004;
7. PIC18F8722 Family Flash Microcontroller Programming Specification, DS39643B, Microchip Technology Inc., 2005;
8. PIC18FXX2/XX8 Flash Microcontroller Programming Specification, DS39576B, Microchip Technology Inc., 2002;
9. Implementing a PID Controller using a PIC 18 MCU, DS00937A, Microchip Technology Inc., 2004;
10. 8-bit Microcontroller Solution, DS39630B, Microchip Technology Inc., 2005;
11. PIC18F8722 Family Flash Microcontroller Programming specification, DS39643B, Microchip Technology Inc., 2005;
12. PIC16F627A/628A/648A Data Sheet, DS40044D, Microchip Technology Inc., 2005;
13. PIC16F62X Data Sheet, DS40300C, Microchip Technology Inc., 2003;
14. MPLAB IDE User's Guide, DS51519A, Microchip Technology Inc., 2005;
15. MPASM Assembler, MPLINK Object Linker, MPLIB Object Librarian User's Guide, DS33014J, Microchip Technology Inc., 2005;
16. MPASM/MPLINK/PIC micro Quick Chart, DS30400G, Microchip Technology Inc., 2005;

Bohdan Borowik February 2010

Kolibri 360 GmbH
Ferdinand-Jühlke-Straße 7
99095 Erfurt, Deutschland
produktsicherheit@kolibri360.de